Dynamic Tourism

Aspects of Tourism

Series Editors: Professor Chris Cooper, *University of Queensland, Ipswich, Australia.*
Dr Michael Hall, *University of Otago, Dunedin, New Zealand.*
Professor Alastair Morrison, *Purdue University, Lafayette, USA.*

Aspects of Tourism is an innovative, multifaceted series which will comprise authoritative reference handbooks on global tourism regions, research volumes, texts and monographs. It is designed to provide readers with the latest thinking on tourism world-wide and in so doing will push back the frontiers of tourism knowledge. The series will also introduce a new generation of international tourism authors writing on leading-edge topics.

The volumes will be authoritative, readable and user-friendly, providing accessible sources for further research. The list will be underpinned by an annual authoritative tourism research volume. Books in the series will be commissioned that probe the relationship between tourism and cognate subject areas such as strategy, development, retailing, sport and environmental studies. The publisher and series editors welcome proposals from writers with projects on these topics.

ASPECTS OF TOURISM 3

Series Editors: Chris Cooper (*University of Queensland, Australia*),
Michael Hall (*University of Otago, New Zealand*)
and Alastair Morrison (*Purdue University, USA*)

Dynamic Tourism

Journeying with Change

Priscilla Boniface

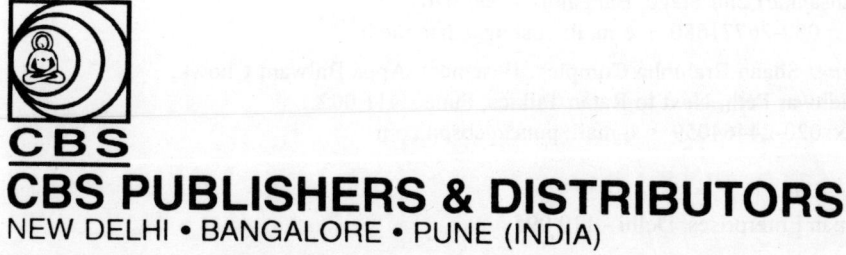

CBS

CBS PUBLISHERS & DISTRIBUTORS

NEW DELHI • BANGALORE • PUNE (INDIA)

CBS Publishers ISBN : 978-81-239-1710-8
Channel View ISBN : 978-1-873150-28-3

First Indian Reprint : 2009

Published by:
S.K. Jain and produced by V.K. Jain for CBS Publishers & Distributors,
4819/XI, 24 Ansari Road, Daryaganj, New Delhi - 110 002, India
e-mail: cbspubs@vsnl.com, cbspubs@airtelmail.in
Website: www.cbspd.com

Branches:
• *Bangalore:* 2975, 17th Cross, K.R. Road,
   Bansankari 2nd Stage, Bangalore - 560 070
   Fax: 080-26771680 • e-mail: cbsbng@dataone.in

• *Pune:* Shaan Brahmha Complex, Basement, Appa Balwant Chowk,
   Budhwar Peth, Next to Ratan Talkies, Pune - 411 002
   Fax: 020-24464059 • e-mail: pune@cbspd.com

Printed at :
Chaman Enterprises, Delhi - 110 095

Contents

List of Illustrations

Acknowledgements

This book very much represents a journey of exploration that is personal, and which is in some ways an odyssey. However, it could not have taken place without others' help and guidance.

I am especially grateful to Andrew Wheatcroft who triggered the motive for 'the trip', discussed my ideas with me before I set off, and whose first suggestions for the book's structure I have adopted and stayed with.

Another person to be singled out for huge gratitude is Mike Stabler for his sensible advice and considerable time given to making improvements to the book near journey's end which were invaluable and instrumental towards my eventually reaching my destination.

Peter Fowler is to be thanked, too, for offering comments at the penultimate staging point before my arrival at the port of handover of the text to the publisher, Mike Grover.

Before entry to that port of arrival, the text went to Anne Sienkewicz. Her skills of editorship and her wider transforming abilities have given the book and its message more clarity, understanding and readability. I would like to express enormous gratitude to Anne for her work in improving the text immeasurably, and in bringing the book to a state of readiness to be led to the voyage terminus of going to press.

Mike Grover has showed the courage of accommodating me among his flotilla. He has demonstrated fortitude as I journeyed onwards, and has given much kind piloting in steering me into the final harbour of completion.

Preface

'Dynamic Tourism' means doing tourism differently. On the one hand, concern is widespread about the potential for damage from tourism. Tourism can have harmful social, cultural and environmental effects. On the other hand, as a tool for change, tourism is widely seen as a chance for social, cultural and economic benefits. Tourism can solve problems, offering new development in some places, regeneration in others. Properly managed, with an eye toward future achievement, tourism can work *for* all, and *overall*, in a dynamic relation between the host societies, the target sites, their visitors and the tourism industry itself.

'Dynamic Tourism' is both the title and the subject of this book. It is a method of management that requires from the outset a fundamental change in concept for everyone concerned. The tourist, first and foremost, as a dominant influence in the industry must change. The tourism industry also must alter its line of thought in an essential way, so that the providers of the tourism product can work in correspondence with their customers. Local and national governments and politicians are all key participants and stakeholders in the process; they too must respond to and cooperate with this dynamic, and so must planners and lawyers, economists and private sector investors. Just as essential are educators and trainers in the field, for they have a vital role to play in forming future tourist industry ventures.

Here is the audience for this book: tourists, governments, planners, tourism industry participants, business people, economists, lawyers and educators. The aim of this study is to persuade this audience of the necessity of their change and involvement, and also to argue for change and involvement in the world at large. Through a combination of theory and example, this book hopes to illustrate the type of change that is necessary and to demonstrate the reasons behind it, while providing ideas and suggestions for suitable and practical action in a wide range of situations. As the title indicates, the central theme of this work is dynamism. Dynamism, fluidity, flexibility and continuous change must become routine and integral to tourism and, moreover, this dynamic tourism should incorporate the essential corresponding elements of lightness and breadth of view.

This study consists of three parts. The first part presents the necessity for change and analyses indications that this change has indeed begun.

Chapter 1

Introduction

Why Change?

Dynamic Tourism is about change, perpetual change over a wide range of components and facets. Its power lies in its speed and flexibility of response, for change is a necessity in tourism. Tourism is not only an industry, confined to a particular sector of an economy or population, it has emerged as a phenomenon that everyone in a society now experiences. We all have our encounters with tourism and are touched by it in some way.

Meanwhile, the industry's method of operation has remained static, beached in a style developed in its infancy. The demand (consumer) side shows a new sophistication that is not matched by the supply (industry) side. No major alteration in approach has yet responded to reflect and accommodate present-day tourism, still less has any alteration been made in anticipation of what tourism will become. Much of the change now occurring in the industry is due to impetus given by the developing Internet and its attendant innovations and opportunities. There has been little of the requisite wide-ranging and strategic thought and pro-active recognition that should drive development of the industry. The main reasons for this industry-wide stagnation and immaturity could be: a blind eye turned by the industry to its own shortcomings; a lack of appreciation for the need and reasons for development; a failure of ideas and imagination.

Why argue for change?

- The impulse to travel – whether a wish to reach a definite destination or a need to escape from a certain somewhere – seems innate in most people. At present, a huge group with this instinct to travel also has the necessary money and time to indulge the impulse.
- New travellers constantly appear, yet established and experienced voyagers, who have already 'seen the sights' are still eager to travel.
- Modern life in our consumer societies demands novelty. We expect new things. We like stimulation. We are increasingly well educated and informed about choices.
- Society is changing rapidly, and tourism within society must reflect and remain in tune with its changes. The need to synchronise with customer tastes is pressing for any industry. It is especially pressing for tourism, since this particular industry bears such a heavy weight of expectation for growth and serves as a motor for economic development.

Tourists, as travellers and as members of society, have changed and are changing; this is the essential point for Dynamic Tourism. Within this dimension of change, in keeping with a constant demand for novelty and a continual replenishment of the new tourist market, it is inevitable that the whole world must eventually be seen as a potential tourist product. Only the entire world can hope to accommodate the eventual needs of tourism. Logically then, even though some populations in some parts of the globe are unable or reluctant to be tourists themselves, it is highly unlikely that they will escape playing the role of host. It is not only travellers who are becoming well versed in the ins and outs of tourism; hosts, too, are acquiring similar knowledge of and sophistication about their industry. Thus the tourist industry, as it delivers its products in these circumstances of continual change, must remain alert, aware of this volatile situation and in pace with it. Not to do so is to risk being marginalised, side-stepped, and done without.

It can be argued that tourism represents a main, perhaps *the* main, material of expression for the developed world in the early twenty-first century. MacCannell (1999: 1) has described the tourist as, 'one of the best models available for modern-man-in-general'. Many features of human life are active in the tourism industry and affect its pursuit, and this human variability accounts for part of the difficulty in maintaining consistent quality in the product. So much of modern society is already reflected in the activity of tourism, for example, our frequent use of consumption as a means of self-definition, our wish for constant, speedy motion, for fresh sensations and experiences. This is the time for tourism to develop as a real reflection of contemporary society, of its thoughts and its manner of operation. If a serious dichotomy develops – a wide division between the demands of the travelling public and what the current travel industry offers – society will abandon tourism for some other, more effective, means of expression and fulfilment.

Tourism belongs to our *free* time. We travel as tourists by our own choice. Such travel can be said, therefore, to reflect our personal essence, since it is a part of life in which we are burdened by fewer constraints and conditioning than in our routine existence. Travel as tourists avoids the duties and influences of everyday life. It gives us the opportunity to do what we really want, or what we think we want, in a concentrated way, without interruption. To aim the product at fulfilling this demanding brief, tourism providers must know us and our needs, and what conditions those needs.

What, then, are the elements that are necessary to satisfy us, and how can we define Dynamic Tourism to embody these elements? One general feature is an attention to catering for our contemporary tastes, outlook and activities where we seek fulfilment through tourism. Yet another necessary feature for Dynamic Tourism must be a mechanism to accommodate burgeoning numbers of tourists. This second necessity presupposes the whole range of environmental requirements – those already in place and those that are still only emerging – as well as protections for landscapes, cultures and communities that must be preserved or sustained as far as possible. Dynamic Tourism also, by definition, must be 'light'; it is critically important for the industry to be able to expand, while still meeting the obligation of serving ever more tourists without harming natural or cultural environments. This 'light touch' must be shown in physical terms, in relation to the Earth, but it must also appear in the approaches that attempt to avoid other harmful impacts on our surroundings, as lightness that allows quick change for more effective response to changing situations. The character and tenets of Dynamic Tourism require lightness of approach and attitude; these are fundamental both for industry practitioners and for tourists themselves. As far back as 1939, Antoine de Saint-Exupéry recommended as much, saying 'He who would travel happily must travel light'.

Moreover, as part of the change understood in Dynamic Tourism, existing tourist destinations that have begun to lose their appeal, whether they are specific places, attractions or hotels, must undergo alteration. Heavy industry has its 'brownfield sites'; these tired and unappealing destinations are the tourism industry's 'brownfield sites'. Their reuse and reconfiguration allows immediate economic and social objectives to be reached in such areas. In the larger context, revitalisation of older sites allows full and efficient use of all the world's tourism resources.

Why Dynamic Tourism?

By now, the outline of a picture begins to emerge, the outline of what is required to produce Dynamic Tourism, a sketch of the needs it must meet. It is necessary that:

- The tourism industry should use fresh approaches to deliver an altered product. The process of change and rejuvenation must keep pace better than in the past with contemporary consumer tastes and concerns. It is necessary to accommodate the increasing numbers of people who are expected to travel should industry growth continue as predicted. It cannot be emphasised enough that the whole range of resources and persons in all sectors of the industry are involved: accommodations, transport systems, destination and attraction areas, tour services, travel agencies, and industry planners. Not only are large-scale providers involved, but many small and medium sized enterprises are also included.
- Tourists, with the encouragement and help of the industry, should bring more deeply into their tourist consumption the impulses and tastes from their everyday life as general consumers and members of society.

To some eyes, the agenda for Dynamic Tourism might appear over radical and unnecessarily disruptive of the *status quo*, even before its particular features and aspects are fully presented and analysed. The arguments for change are:

- Tourism, as it stands, is already not working very satisfactorily for many involved in it, whether as tourist or provider.
- The potential of the industry remains unseen or unaccepted by many people, and so remains largely undeveloped.

In its definition of tourism, the Commission of the European Communities (1995: 24) observes that: 'Very few activities are as directly involved in the process of transforming and developing modern societies as tourism', and continues that, 'tourism is the medium *par excellence* for bringing people close together'. Tourism's strength and potential as an engine of social transformation are seen here, but tourism is also shown as a mechanism that can provide possibilities for individual personal development, and can also generate opportunities for individuals to meet.

People who provide the tourism product must recognise more clearly and respond more acutely to the needs and wishes of their customers, both as individuals and as members of the greater society. They should identify how tourism can serve these requirements while also meeting the needs of the industry itself. This is an industry that is still in its early phase. Tourism has yet to 'grow up'; it is falling behind in its markets. Essentially, the industry has remained at the 'package holiday' stage, as it was developed by Thomas Cook in the nineteenth century. Turner and Ash (1975: 59) remark on the benefits and defects of that approach, saying, 'This new standardisation was double-edged in effect; on the one hand, it meant greater comfort and convenience and less need for decision making on the part of the individual tourist; on the other hand it meant a decrease in the elements of real novelty and adventure in tourism'.

Inexperienced tourists are attracted, of course, to this kind of package holiday, where they may avoid much individual effort, decision and responsibility. There will always be such novice travellers, since new tourist markets are continually emerging. On the other hand there is also a large pool of extremely experienced holidaymakers. Such groups of seasoned voyagers will find the 'nannying' aspect of package tours irksome. The removal of individual choice is a disadvantage for them; they want something more suited to their level of sophistication and knowledge. This tendency in tourists to become more independent and sophisticated is typified by the Japanese. At first, they journeyed abroad on holiday in intensely organised groups. Now, many Japanese holiday on their own, as individuals, couples, or in small clusters of friends.

Observation of such developments suggests that more of us want greater flexibility in travelling than is permitted by the traditional package holiday. We are seeking greater personal choice and more room to allow for individually driven fulfilment. Moreover, as general consumers, we increasingly expect high quality and sophistication in the products we buy. Furthermore, as global citizens we have been forced to accept the fact that our world is finite and sensitive; it is a vulnerable resource. The combination of all these elements points to the identification of more tourists who fall within the categories of adventurer, discoverer, pilgrim, explorer, culture vulture, eco-tourist and environmentally sensitive voyager. Their values ought to influence the available models of contemporary tourism. Already there has been a transformation: what was once seen as 'alternative' and 'minority' has come to be mainstream in today's tourism. The classic definitions of tourism have lost much of their appeal, because well-seasoned travellers want more than routine, customised approaches. As Turner and Ash (1975: 138) reflect in commenting on historic monuments, 'it is arguable that sites that have become institutionalised tourist sights are no longer worth seeing'.

There seems to be a lag in awareness of the significance of this change in the tourism industry. In 1995, Barrett (1995: 64) reported that government statistics for the United Kingdom had revealed that, over a five-year period, independent foreign travel had increased by more than 32%, whereas package travel had shown an increase of only 20%. The director of the independent firm WEXAS was quoted as claiming that, 'the mainstream travel trade is failing to cope with the more sophisticated demands of Nineties travellers'. Argyle (1996: 264–65) describes Pearce as producing five traveller types in 1982 through the assistance of multi-dimensional scaling: Exploitative, Pleasure First, High Contact, Environmental and Spiritual. Three of these types are likely to demand a Dynamic Tourism kind of product. The Exploitative type, exemplified by business travellers and jet setters, clearly includes the target traits of independence and experience. Only the Pleasure First tourist is likely to fall within the category of the stereotypical 'package travel' holidaymaker.

Social Influences for Change

As certain key characteristics become apparent in society in general, they are also revealed as present among consumers of tourism. It has become a worldwide trend, revealed in the overall reduction of Communism, that societies are less controlled and regulated than in the recent past. There is a general move in the direction of reducing hierarchies and increasing social equality. The political climate is moving toward styles that place more emphasis on individual responsibility and freedom of choice, with Capitalist financial styles in a more prominent position.

Linked to this change in climate, there is a global unity of attitude on a range of matters from the environment through to finance. As technology makes information more readily available, the development of common patterns of style, conduct and attitude is hastened. This technologically driven speed of communication now means that changes of taste and opinion can occur and spread rapidly. Increased sophistication and technological capacity ease the process of change; more radical transformations can occur with less radical difficulty. Thus one of the key concepts of Dynamic Tourism, that tourism be able to operate successfully throughout the entire world, is nearer to meeting its necessary prerequisites of globally concerted aim and activity.

Recognising the existence of this new trend is not enough. Its existence and the changes necessary to facilitate it carry implications and make demands on the tourism industry. Mulgan (1997: 50–51) argues that the fact that we are living together as a society, and that we need to do so, requires us to restrain our individual freedom to some extent, and to develop considerable management structures. Plog (1994: 50) specifically recommends, 'To protect the world for tourism, and protect the world from tourism, a common set of goals, evolving into specific plans for different regions, is desperately needed'. Of course, the critical aspect in defining tourism frameworks within the principles of Dynamic Tourism is to ensure that the frameworks are of a style and quality to allow flexibility for responsiveness and change.

Take museums: many of them are good examples of traditional tourist attractions that need to adapt to change. To remain relevant and attractive to consumers, museums must demonstrate their customer orientation; the traditional position of a haughty superiority of knowledge is no longer possible. Many museums still need to learn to communicate with a wider audience, in a style that has more popular appeal. In describing the democratisation of the museum world, MacDonald and Alsford have emphasised the need to offer the public, 'increased opportunities for personalised and self-directed learning along pathways set by individual interests'. They continue, more generally, that:

> Democratisation means greater participation by the community in the content and direction of museums. ... Society places greater value today on accountability and public involvement with public institutions. If museums are to remain viable service institutions, they cannot keep aloof ... but must find their own ways to respond to changing social demands. (MacDonald & Alsford, 1989: 40)

The 'greying' of the population, as the proportion of the elderly grows, has a critical impact on tourism. Today's elderly are not the same as those of yesteryear. Generally speaking, they are affluent, fit, ready for adventure and exploration, and keen for personal development. In the United States (US), a major market of tourism consumers, 'the fifty-plus age group will make up 38 per cent of the population and hold 75 per cent of the nation's wealth' (van Harssel, reporting Wolfe in 1990). Using US Travel Data Center figures from 1990, van Harssel predicts that travel by the US mature market 'will significantly increase'. He continues with a comment based on Forbes (1987: 17–21), 'Today's older Americans are significantly different from earlier generations of seniors. They are better educated, more active and more vocal than any previous senior generation. Advanced years are now seen as a time of challenge' (van Harssel, 1994: 364). Handy, discussing Third Agers and seeing them as having little opportunity for earned income from work, predicts that 'They will ... begin to find that there are satisfactions and achievements which cannot be measured by money'. He sees the group as engaging in 'study work'. Since there are so many of them, 'they will be noticed'. Handy (1994: 232) continues his description of Third Agers with the comment, 'They will not be old, as we used to think of old'.

Society is ageing. It is also becoming more urban. The day-to-day life of about half the world's population is conducted within an urban milieu. This is a fundamental dimension of the definition of contemporary and future tourism products. Essentially, the matter is this: as human beings, we are most whole and content when we are in a balance of opposing and complementary factors. As tourists, we usually seek difference and escape from our routine way of life. Thus, urban dwellers on holiday are drawn to things they cannot easily find in the city, just as residents of the countryside relish city holidays for their charming novelty. Another inference we may draw is that, for maximum consumer satisfaction, each holiday should offer a balance of contrasting experiences and sensations. Here again, we must stress that our core requirement is to satisfactorily identify the needs of tourists. We must re-emphasise the message that the world's full range and diversity can and should be represented in tourism. This means that a full complement of types of venues, accommodation and transport should be offered as appealing to tourists, that they should all be formulated in new and different combinations and offered on varying occasions with an eye toward fresh circumstances. Furthermore, we must stress that, in the course of this general attempt to encompass all possibilities, no matter which of these changing dimensions is included in the tourism entity, it must represent high quality in its category.

Within the field of tourist product elements, there are certain critical contrasts, where differences are pushed to the extreme; the elements of people, modernity, overtly man-made monuments and manufactured entities compare vividly with the elements of peace, quiet, traditionalism and nature. Stereotypically, we divide them between the city and the countryside. If the theory holds true, then the implications for planning for both distinct areas, in tourist terms and in societal terms, are considerable.

Already in 1967, Marshall McLuhan and colleagues (McLuhan *et al.*, 1967) were of the opinion that, 'The city no longer exists except as a cultural ghost for tourists. Any highway eatery with its TV set, newspaper, and magazine is as cosmopolitan as New York or Paris'. They were suggesting that, with the introduction of modern communications, cities had lost their distinctive edge of difference in comparison with the rest of the world. For the appeal of the city as tourist destination or residence to be vaunted and used, the actual key elements of attraction must be carefully identified and provided for. The same is true, of course, of the countryside; careful definition of what visitors expect and want is needed.

Clearly, certain definite cultural monuments and formal attractions that are not encountered in rural areas are counted among the elements of urban appeal. However, this very strength in original appeal can be turned against these monuments, when they are formulated and presented to cater to great numbers of tourists, in the expectation of meeting presupposed tastes. As tourist destinations become institutionalised, they lose their quality of fluidity; Turner and Ash (1975: 138) say as much in their earlier quote. The possibility of alteration is gone; monuments become static and removed from their everyday, participatory existence within the city. Cities must demonstrate the presence of human qualities: nature, flexibility, lack of rigidity, order and symmetry; there must be a reduction in the hardness associated with modernity and technology. These imperatives have led to the appearance within cities of the contrasting elements of parks and greenery. As an example of a formal venue, we find these impulses represented in the organically shaped curves of the Canadian Museum of Civilization. Chosen to be the museum's architect in 1983, Douglas J. Cardinal deliberately called upon nature as his inspiration. MacDonald and Alsford (1989: 13) comment on his approach, saying, 'his style [is] humanistic and uplifting; he places human needs at the heart of the design process'. These are among the qualities that fulfilled Cardinal's commission to make the Museum 'people friendly' (ibid. p. 18)

It may be easy to pinpoint certain great monuments as elements of urban appeal, but perhaps the most under-recognised attractions are people themselves. Everyone likes to watch other people, tourists watch the residents and residents watch the tourists – and each other. Look at the people sitting in sidewalk seats at cafés, watching the world go by. The roots of such behaviour are in the *flânerie* so dear to the hearts of Parisians of the Belle Époque. For them were built the covered shopping galleries, forerunners of today's shopping malls. These arcades allowed city dwellers a pleasant stroll, sheltered from the elements. Though long undervalued as an issue, the matter of adequate provision of public space is now being given greater weight. One such voice is that of architect, urban planner and commentator Richard Rogers (1997: 152). People watching, with its dimensions of constant change and serendipity, partakes of the essence of Dynamic Tourism.

Often rural and undeveloped areas provide a contrasting appeal that balances the appeal of urban areas. The countryside delivers the attraction of a less complicated basic environment and a way of living that is oriented toward community and village life. The agricultural and seasonal cycles of country life focus attention on the essential continuity of nature. These features, seen as remnants of a simpler lifestyle, cannot be experienced in the rush and modernity of the urban cityscape. Such dimensions of life have a special attraction for people who love security. So often, the invitation of beach and seaside holidays also is connected to an essential longing for a return to a simpler existence. Half-remembered, romanticised images of childhood trips are now perceived as encapsulating essential values, carefree times without the attendant troubles of modern, adult life. The countryside or coast seem to offer peace, quiet, a chance to get away from it all, away from the complexities of contemporary city life.

One key, different feature that remote, rural areas can provide, is the prospect of adventure, the possibility of testing individual capacities for enduring hardship and extremity. Such testing occurs in the course of dealing with harsh terrain, extremes of climate, and the lack of modern amenities. Urban dwellers are given the opportunity of studying and experiencing those aspects of life that are peculiar to the countryside, but rare and unfamiliar in the city. The country offers unfamiliar plants and wildlife, caves, forests, streams. There is the chance to observe farmers at work. There is also the primary quality of rural life: a chance to experience and inhabit outdoor space. As urban populations increase and become denser, an increase in the power of attraction of outdoor space may be expected.

Rural areas may also be seen to offer the element of spirituality, perhaps due in part, to the simple absence of urban rush and stress. As society becomes more secular in overt terms, people seem to need some form of spirituality, perhaps as an antidote to the technocratic nature of much of our modern world. Spirituality may take the form of a retreat, which is always easier in the relative remoteness of the countryside, with fewer potential interruptions than in the city. Such a retreat may be formalised and thus be the main structure of a holiday, or it may be informal, simply felt as an element of escape from the everyday life into the holiday existence. Often, in the interests of spirituality and personal development, people expect their travel event to include some dimension of suffering, or a difficult demand of purification and enlightenment. Some sports events, treks and activities offer this dimension of personal testing and thus serve the extra, complementary role while providing exercise and physical exploration.

While exploring the theme of opposition and balance, we must not forget, however, that for those people who are so inclined, the city can be an equal source of spiritual refreshment and mental stimulation, with its congregations of people, ideas and major monuments. Visitors may come to the countryside or the city, whichever they personally regard as having the capacity to refresh their souls and extend their beings. We may generally say that few travel experiences fail in delivering some degree of personal change, hence the old cliché that 'travel broadens the mind'.

Many people now live modern, materialistic existences that are centred on technology through such lengths of time as to be bored and sated by them. For this reason, and because of the contemporary perception that a reduction in the standards of living of the developed world is likely to be necessary for the survival of the planet, a new sense is developing – a yearning to attempt a new way of life with reduced emphasis on materialism. Yet another perspective on our life-long, routine consumption of material goods is the view that we no longer need more physical items. As increasingly mature consumers, we have had our fill of owning things. We do not even want the trouble and responsibility of ownership.

This idea of being 'thingless' is familiar; it relates to a central tenet of Buddhism, a religion that not only has adherents in its traditional heartlands of the East but nowadays also engages attention in the West. This outlook was explained by Daisetz Suzuki (1993: 120), the leading Zen Buddhist authority: 'The desire to possess is considered by Buddhism to be one of the worst passions with which mortals are apt to be obsessed'. It seems logical to refer to Zen Buddhism here, since its philosophy contains many elements that are suggested here to be fundamentally integral to the Dynamic Tourism concept. Dynamic Tourism presents itself in its fledgling existence and argues for the necessary institution of a new perspective characterised by attention to basic and elemental things. Similarly, Suzuki (1993: 88) explains 'The object of Zen discipline consists in acquiring a new viewpoint for looking into the essence of things'.

Yet another idea about our society suggests that we have reached such a level of consumer sophistication that we want to pick and mix the elements we consume in a carefully defined way, across a huge range of choices. We do not want any constraints or expense of ownership to limit our response to our choices. The general trajectory of this proposition culminates in a concept with special significance for the area of tourism, especially when linked to the foregoing Buddhist perspective. Within the context of the search for experiences, consuming them and piling them up, only the most essential and important material goods will be collected and kept. These special items will be acquired for their contribution to basic sustenance or because of their special significance, their value as talismans of experience. Within the definition of Dynamic Tourism, we want to 'travel light'.

Of course, this concept of self-restraint relies upon and demonstrates our standard of living and growth of expertise as consumers and tourists in the developed world. Were our basic needs not being met, as indeed they are not for many people in parts of the Third World, we might not choose to play at being nomads or to exercise the luxury of chasing and gathering experiences rather than material possessions. One common element in all tourism is that it is engaged in only on a discretionary basis, to serve higher needs, such as self-fulfilment and spiritual enrichment. Tourism is not a basic survival necessity. The virtue to the search for experience, which is proposed as a key feature of Dynamic Tourism, is that some elements cost nothing or very little, and so are available to a wider public. The critical thing for all tourists, rich or poor, seeking experiences and encounters is information, so that they can make knowledgeable selections.

These recent remarks outline the essential premise that, as members of society and tourists, we are showing and will continue to show, more of the characteristics of the nomad. This does not mean that we are, can be, or would want to be nomads in the old sense. A more appropriate name for tourists who display these current tendencies is Neo-Nomads. For certain parts of society (and thus for tourists), there seems to be a real attraction in the idea of reducing commitments and wandering between situations where extreme contrasts may be encountered, making choices, selecting and experiencing necessities, without making major efforts or carting about great numbers of impedimenta. The new 'greys' seem as likely to display these leanings as do today's young people who tour Europe on InterRail, or circumnavigate the globe with a rucksack as their only baggage. In this style of tourism, an impulse can take the shape of a pilgrimage and become a spiritual search, with a possible particular connection with aspects of the philosophy and style of Zen Buddhism that we have already described.

Mulgan (1997: 47), founder of the think-tank Demos, links the activities of nomadism to self-sufficiency and freedom – a perception that fits with what we have seen of our modern-day nomadic tourist, who is travelling by choice, not necessity, with resources drawn from a foundation of knowledge, capacity and maturity. Harking back to the nomads of old, Mulgan notes that they 'used more of their mental capacities and arguably had a fuller life than their successors' (ibid. p. 48), an expanded nomadic model that may serve as a template for today's Neo-Nomads. Mulgan goes on to assert that the idea of nomadic mobility is surprisingly prominent in our culture.

However, for Neo-Nomads to find their wanderings feasible and rewarding in modern society, they absolutely need reliable information. In fact, the accessibility of large amounts of data is so essential to this travel style that the easy availability of information may have caused the resurgence of interest in nomadic life. In an absence of data, or without easy access to information, these wanderers would simply never have re-emerged as figures of contemporary society. It was Marshall McLuhan (1970: 43) who made the direct connection of these ideas with the words, 'Bless the Electric Return to the Tribal Palaeolithic Age. To the World of the Hunter after the Neolithic Planter'. McLuhan thus indicates a return of the nomadic non-acquisitor and suggests that information technology is a prerequisite of this renewal.

The revolution in easy access to information was still very much a novelty when McLuhan wrote the above passages. In the same text, he reflected in a similar vein, 'In the Age of Information man the foodgatherer returned as man the factgatherer ... Circuitry is the end of the neolithic age'. (ibid. p. 33) Looking back on the late twentieth century, it is indeed easy to see that it was indeed an era of information. The revolution heralded by McLuhan has largely come to pass, though developments in information technology continue apace, of course, and information networks continue to extend.

Information is not an end in itself. Networks and connections exist in order to exchange gathered information, but they can also enable new information to be obtained and thus advance learning and knowledge. McLuhan recommended, 'In the global village of *continuous learning* and of *total participation* in the human *dialogue*, the problem of settlement is to *extend consciousness* itself and to maximise the opportunities of *learning*' (ibid. p. 41).

As tourism continually brings different people and populations together in new encounters, it becomes a medium *par excellence* for progress in making useful and stimulating discoveries. We need to have this opportunity presented to us. Peters (1994: 174) delivers, as an unconscious by-product, a blue print of justification and promotion for what tourism can provide, as it transforms us into us host and/or guest, in new situations. 'It's simple ... We either get used to thinking about the subtle processes of learning and sharing knowledge in dispersed, transient networks. Or we perish.' Yet another recommendation from Peters (made in the context of desirable characteristics for staff appointments) is relevant for providers in the tourism industry: 'Hire curious people' (ibid. p. 200).

Here it is necessary to pause and re-establish the connection between the concepts of Dynamic Tourism, with its characteristic of travelling light, free of the heavy baggage of intensification and unnecessary pre-ordination, and the precepts of Zen Buddhism. Christmas Humphreys (1990: 180) attributes to Zen the purpose that it 'is to pass beyond the intellect'. He adds, 'Zen ... strives to KNOW' and continues, 'For this a new faculty is needed, the power of im-mediate perception, the intuitive awareness which comes when the perceiver and the perceived are merged as one'. The path of development suggested is from information and its direct acquisition toward a more instinctive absorption of data and knowledge, thus leading to the use of this acquired data in a creative fashion and towards a creative purpose. MacDonald and Alsford signal this transition as they comment on the nature of museum collections of artefacts:

> Their importance has not lessened, though it might seem so in the light of new activities of the museums that are less *directly* dependent on the display of original objects. If anything, it is greater, thanks to the fuller appreciation of the many ways they may be used to satisfy diverse user-needs: as national treasures, as icons in the information system, as reference points in our reality grids, as masters from which replicas or images may be disseminated to a larger audience, as sources of inspiration giving rise to new creative expressions. (MacDonald &Alsford, 1989: 64)

The proliferation of data in the twentieth century well justifies dubbing it the 'age of information'. Information is needed as a basis of embarkation, as one century ends and another begins. Richard Rogers (1997: 151) called for us to develop toward the future through the exercise of creative imagination urging the development of a 'creative society'.

The developments heralded in our age of information and changes in our society as we move toward the future have a key bearing upon tourism. In 1997 Mulgan spoke with direct relevance to this point, bearing out many of the wide aspects that characterise Dynamic Tourism and its precepts:

> The rise of exchange has left us with an unprecedented density of institutions, cultures and people whose *raison d'être* is exchange and transformation, taking something and turning it into something else, as opposed to concern for land, territory and direct engagement either with material things of with others as people to be cared for. It has shifted the centre of gravity of societies away from those occupations which favour continuity, into occupations which favour change, unpredictability, spontaneity, innovation, creativity. (Mulgan, 1997: 90)

Tourism's Response

Because of the state of flux and change in contemporary society, it is essential for tourism to embrace provisionality, fluidity, flexibility and serendipity. These will be shown as integral to Dynamic Tourism. The recipe may seem dangerous and fraught with uncertainty, and to an extent it is, but the other option open to the tourist industry, that of failing to respond to the prevailing society and consumer taste and style, is more perilous still. The methods proposed under the name of Dynamic Tourism demand a de-construction of the current industrial approach, and a subsequent re-construction in a fresh form.

In general, tourists wish to watch, experience and evaluate lifestyles and ways of handling life. This is the basic role of tourism, and this function must be seen as important and crucial for society, thus endowing tourism with a critical, useful purpose that far exceeds its disclaimed position as only a sector of society and one that is superficial. Dynamic Tourism tries to offer a way for the potential of tourism to be realised. How can these needs be attained through the practices and products defined as Dynamic Tourism? Detailed examples of particular recommended activities will be provided in ensuing chapters; in this conceptual introduction, things will be put in general terms.

The essence of dynamism is not merely to change, but to be in a constant state of change. By definition then, this constant change is an essential element of Dynamic Tourism. It presupposes rapid alteration, and thus a provisional nature becomes an on-going feature of its practice. This means that any given tourism management structure and edifice of accomplishment, and the tourism product they present, must be seen only as the best current solutions. They are temporary positions, subject to change if wants or circumstances alter, or if improved models are found or developed. Dynamic Tourism is formulated and delivered in response to the needs of society and of tourists, as these needs are best perceived and understood. Among its aims are the imperative of sustaining environments and cultures, as well as of enabling hosts and operators to be adequately paid for their work. However, it should be emphasised that the means for payment may need to be altered; remuneration may arrive by less direct avenues than formerly was the case. This is true because, as attractions are less formal and pre-defined, and also spread over a greater distance, there will undoubtedly be some loss, at least on occasion, of direct opportunities of generating income from tourist visits.

Therefore, the primary revenue sources to be explored will lie in those items (food, souvenirs and guidebooks, for example) that are sold alongside attractions as adjuncts to their desirability. This approach represents a shift in emphasis, rather than a revolution, since making money from support activities rather than from the 'gate' or from the original site of interest, is already well practised. This is an example of how Dynamic Tourism requires people in the tourist industry to rethink and change their ideas and endeavours. To ensure present success and future viability, the guiding principle of the industry should be based on a strong understanding of the needs of society and tourists, and on the ambition to meet those needs in optimum terms.

Today's consumers and travellers have become sophisticated and knowing. People who cater to tourists should not think that their customers can be duped, nor that they will not notice inadequacies in service and product, nor again that they will accept less than the full perspective. Donald Horne (1992: 201) draws attention to gaps and unsuitabilities in some museums' representations of the past, 'when whole classes of people, even majorities, are ignored, or may be present only in ways not related to their own views of themselves'. He asserts that 'These silences in the public culture, transposed into the silences of tourism, provide visitors with a national past of the country they are visiting in which practically all the characters are missing, and in which there is virtually nothing of the country as it is now'. The example could be transferred to the wider tourist scene; whenever there is a choice to be made, we need the whole story to help us choose suitably and responsibly.

It is clear that education and training are central to achieving a general understanding that change is required in tourism. Needs and reasons for change must be communicated; the tourism education sector must take the lead in seeing that future leaders and participants in the industry are well informed. Tourists, also, should understand their position; they too require the help of an adequate supply of expert, up-to-date information. Tourists hold great power; they are the people to whom the industry must cater, the ones whose voices must be heard. There should be help available to make them conscious of their responsibility to be guided by more than the pursuit of their own individual experience in making choices in tourism, rewarding as that individual experience might be. Their responsibility calls for them to decide what is most suitable for society. As the mature travellers that many of us now are, we usually know what we ought to do for the common good; but for us actually to do it, we may need to be told so with greater emphasis.

Chapter 2

Signs of Change

Already *En Route*

Dynamic Tourism responds to the changing needs of society and tourists. As a response, it follows on needs that already exist, or at least can be seen emerging. In some quarters, contemporary requirements are recognised, and the alteration of effort to serve them has already begun. Signs of change can already be seen.

Chapter 1 outlined the changes in society and in tourists that call for a reaction from the tourism industry. The suggestion is that, while tourists have changed and continue to change, the industry is being left behind, because it has altered much less than its customer base. The essential changes that demand adaptation are discussed, and it is asserted that these changes are produced by our increasing responsibility and interdependence as global citizens, as well as our increasing independence, freedom and sophistication as individual travellers. To exemplify the style and manner of approach and action dictated by Dynamic Tourism, this second chapter will portray the changes that are occurring in society and amongst individuals, and that are present or emerging in the tourist industry itself, as well as in its partners and stakeholders.

Certain key features stand out, some of them interesting and surprising. These are they. At the same time that the Western World uses technology and espouses modernity and the cult of novelty, we also pursue a taste for nature and immerse ourselves in it. Moreover, we are constantly rediscovering, evaluating and deploying old ways and methods within our contemporary lives. Though our society is becoming increasingly secular, with a certain rejection of formal religious practices, we are yet engaged in an active search for spiritual immersion and expression. In the world at large, hierarchies and dictatorships are waning. We recognise ourselves as truly equal with each other, with no single philosophy holding a monopoly on the truth. We expect our standard of living to continue to rise. However, to be realistic, if everyone in the world is to enjoy a reasonable level of existence, and if future generations are to have a fair chance of life, the developed world must accept a reduction in levels of material advance. Thus the interests of the entire planet may require that industry and its potential are not fully realised.

We have huge resources, but for the most part they are finite. Their exploitation sometimes has harmful effects. They should not be used recklessly and, wherever feasible, as much as possible should be recycled and reused. We are so accustomed to easy gratification with material goods, that we have reached the point where either we need and desire no additional physical possessions, or else we want quicker and quicker 'fixes' of novelty to sustain our interest. We can obtain information easily and rapidly, but is that information then seen as only a starting point toward new ideas and initiative, development and fulfilment? Do we put the information to full use? Our lives are complex, cluttered with goods and activities, so we relish simplicity as a change in itself, and we find it unusual, interesting and appealing. Perhaps much of what has just been said indicates the need for balance and wholeness in our lives. Perhaps it shows our implicit recognition of the benefits that come from experiencing contrast and variety. There is a dominant, essential role for the experiential factor to play in our lives. At this juncture, where physical goods have become routine and easily accessible, we are looking for direct experience, as a quality aspirational feature.

This is the background of events. Their relevance is clear. At this point, it is proposed to introduce more detailed evidence. In particular, indicative dimensions in society that are related to tourism will be identified and considered, then tourism itself will be studied.

Information, creativity and experience

In general, as the millennium approached, people sought to develop in the realms of materialism and science. Chapter 1 cited the projection by Rogers that we are now entering an age whose critical aspect will not be information, but the creative and imaginative use of information. Rogers' predicted creative approach is already found among consumers in the realm of tourism. With their sophistication as travellers and the fortification of knowledge that they possess through participating in the information era, voyagers are increasingly well equipped to make more of their own choices, and they are increasingly keen to do so. They are ready and eager to make their own individual formulations from the array of destinations, attractions and types of transport and accommodation available. The key information sources for them are guidebooks, specialist magazines, newspaper editorials, television and the Internet, rather than perhaps only brochures from tour operators with their essentially rudimentary data, so often subjectively upbeat rather than strictly objective.

Professor Peter Cochrane (1995: 7), in considering the potential of virtual reality to allow us 'to enter the real world from a distance', has asserted that, 'in about the first decade of the new century', there will be a 'transition from the information society to the experience society. It will be about "being there"'. His remark highlights the contemporary transformation into a general societal motive of our wish to travel from a position of virtual reality to concrete experience. Within these remarks we may perceive three important aspects: travel, virtual reality and experience. Travel has long been with us. Virtual reality is well upon us. Experience is already being revealed in tourism to the extent that it may now be the key objective of today's traveller. We are 'shopping' for experiences. We want to try things out and empathise with other people's lives, and to sample new activities, in order to enliven and enrich our own existences.

Spirituality and nourishment: holism, nature and simplicity

The search for new sensations, for greater depth and satisfaction in existence, manifests itself in other ways. It is part of our hunt for contrast and counterbalance, and part of our search for the spiritual dimension in our world. An example of one person who prizes the experience of heightened spirituality is Japanese mail-order multimillionaire Katsuhiko Yazaki who, 'after a spiritual experience while meditating in a Zen temple', began to take a larger interest in the future, eventually establishing the Future Generations Alliance Foundation. He also authored a book, in which he is reported by Albery (1994: 18) to say, 'We cannot find true meaning in life by occupying spacious residences. At some point people will need to raise their desires to a higher level'. His allusion is that we have material possessions but, for fulfilment, we need something beyond them.

Professor Christopher Frayling (1995: 8), a cultural historian, presents yet another viewpoint. In evaluating the notion of a contemporary return of interest in the High Middle Ages, he attributes this idea to being part of a search for *nourishment*, for roots Frayling's remark evokes that aspect of 'returning to the past' in our lives. He finds this notion to look back in time to be useful in our search for personal enrichment. The revival of interest in ancient approaches and ideas contrasts with modern ways, and this encompassing of opposing lifestyles offers the desired experience of 'wholeness'.

Perhaps, in some ways, Japan epitomises the notion of wholeness and does so in an aspect that seems to come very close to our contemporary needs. This country seems to perform the delicate balancing act between concentrating on what is new and modern, and still venerating and drawing strength from elements of tradition and the past. A focus on simplicity is integral to Japanese culture, where it is manifest in a number of dimensions. The traditional verse form of haiku is a good example, or again, there is the Japanese reverence for nature, as demonstrated by the attention they give to viewing spring cherry blossoms. Consider how Japanese gardens, of which the *stroll garden* is one type, provide different landscapes and viewer experiences within their perimeters.

Yet another example of the return to ancient wisdom is the use of remedies based on herbs and flowers in alternative therapies. This phenomenon is not a minority endeavour. *Vogue* magazine, hardly a voice of the counter culture, has proclaimed (in an article by Mimi Spencer), 'Cranky no more, alternative health is becoming the smart option for modern women'. Furthermore, 'In 1984, most upwardly mobile mouths would have smirked at an acupuncturist's needle and laughed in the face of a yoga posture, even as a notion for the future' (Spencer, 1995: 108). As the *Vogue* piece makes clear, a key attraction of alternative therapies is that they offer personal control and responsibility for oneself. The article's conclusion, offered by top model Tatjana Patitz, is that 'Technology is destroying us. ... Nature has a remedy for every illness, an answer for every problem' (ibid. p. 112).

The East often serves as the source of alternatives and of help in spiritual searches. The recent upsurge of interest in Buddhism in the West is a proclamation. Perhaps we may find the roots of today's modern movement in the 'flower power' initiative, lifestyle and alternativism that emerged in the late 1960s. This alternative movement is exemplified in the Beatles following of the Maharishi, albeit only for a while as an entire group, and by hippies setting trail for Kathmandu.

Nature appeals to the tastes of modern society. Flowers, trees, grass and water are especially prized key attractions. The late Diana, Princess of Wales, was allegedly a devotee of Bach Flower Remedies (Alexander 1995: 32). The appeal of gardens for humanity has continued through the ages; it is extremely strong in our present-day lives. For example, one National Trust property in Sissinghurst in Kent, the former home of writers Vita Sackville-West and Nigel Nicholson, is a visitor attraction, with its garden the key feature of its appeal. Admittance is charged to the property, and such is the popularity of the site that timed ticketing is necessary.

Why do gardens attract us so much? Surely their charm is due to the following elements:

- Gardens offer stimulation and knowledge about technical horticulture, the design of garden layout, for example, or the types of plants to use in particular situations.
- They provide open spaces for physical recreation and relaxation.
- They serve as picturesque venues and backdrops for gala events.
- They offer variety within themselves, as well as across the range of garden possibilities.
- They even seem to extend into the realm of aromatherapy through the charm of their varied scents.

Taken all together, the overall appeal of gardens seems greater than its parts; it extends into the realm of spiritual refreshment. Gardens serve people's deeper needs. They are exemplary models of many of Dynamic Tourism's tenets, and will be promoted heavily throughout this book as great tourist attractions that merit increased attention and development.

Gardens have a range of capacities, while also being very versatile and amenable to change.

Another indication that today's society is seeking to meet a spiritual need was the chart-topping global popularity in the 1990s of an album of Gregorian chant recorded by Benedictine Monks at the Spanish monastery of Santo Domingo de Silos in 1973. As one of the monks commented (Lamb, 1995: 22), 'This is music which touches the deepest part of the soul and transports one to another world'. In a more prosaic vein, but still transmitting nearly the same essential message, an EMI representative (Penman 1994: 16) remarked that 'almost anyone seems able to use the music and tune out their worries and relax'. Writing in *The European*, in an attempt to explain the general revealing feature of appeal, Holland employed no nonsense terms, with a strong significance, 'The monastic orders have always held a fascination for their more worldly brethren' (Holland, 1994: 3). Taking a larger view at the overall aspect, he noted:

> Marshall McLuhan's global village had achieved its purest form yet. 21st century technology was being used to broadcast an ethos of silent solitary meditation. The phenomenon is strange by any standards. Old and new, religious and secular, technology and soul: the fashionability of the spiritual. (Holland, 1994: 2)

Holland picks up on the contrasts that correspond to the essential need for wholeness that is examined in this book. Santo Domingo de Silos became a major venue for tourism following upon this musical success. As Penman (1994: 16) remarks, 'There is appalling irony in the fact that the monks have inadvertently opened the door on the material world'. The monks had joined secular aspects of our universe to their own cloistered existence, while providing a spiritual dimension to the outer world.

The compact interactive disc series 'Europe facing the past' is produced by the Belgian sector of the European PACT network with Phillips. The series of discs is defined to provide Themes, a Guided Tour and Direct Access. It is significant that the first topic chosen for a disc in the series was Cistercian Architecture. Presumably this choice was carefully made, in the belief that the subject has considerable consumer interest and appeal. In this topic, spirituality and religion are central concerns. Of further significance is the fact that the characteristic features of the Cistercian order were simplicity and rigour.

For some tastes, indeed, the Cistercian ideas, buildings and approaches are not attractive; they seem inadequately ornate, insufficiently flamboyant. Yet it is clear that they do exercise a great deal of current appeal, perhaps because they offer a welcome antidote to present-day complexity. For example, in 1999/2000 the National Trust's most popular charged-for venue was Fountains Abbey in conjunction with the gardens of Studley

Royal; and this complex is also a World Heritage site (Boniface, 1996: 112–13). This example underlines the appeal of gardens and the Cistercian tradition. Sissinghurst, which was mentioned previously, is another of the Trust's top paying attractions: in 1999/2000 it was ninth in the 'league'.

In France, the Cistercian Fonteney Abbey in Burgundy is also designated a World Heritage site. Fonteney has demonstrated considerable dynamism in its time: it has evolved from an active place of religious life and worship to a paper mill, and is now restored and serving as a tourist venue. The nine hundredth anniversary of the founding of the Cistercian Order in 1998 attracted a prolonged and high-profile celebration in Burgundy. This celebration was promoted in the summer 1998 magazine of the Société des Autoroutes Paris-Rhin-Rhône' and was marked by an array of visitor events.

Transport, 'green-ness', and walking, in particular

We know we should use cars more sparingly, it is part of our identity as responsible world citizens. We experience the pressure of environmental concerns and we feel the logic that operates in line with our consideration of simpler, 'lighter' ways from the past, ways still followed by less 'developed' peoples. We realise that we should at least try to modify our fixation with individual car ownership and to generally use automobiles more sparingly. The situation is difficult because many twentieth-century freedoms were made possible by the automobile. Ostensibly, at least, it serves many current social needs and pre-dispositions, including those of tourism. It can seem that the freedom, spontaneity, chance and option of choice, which are the promise of Dynamic Tourism in response to society's restrictions, are met through individual vehicle use and ownership. However, the other part of the equation is that, in our maturity, with our feelings of responsibility for the environment, we need to look for alternative ways of transportation, rather than employing a multitude of individual cars, and most especially those that run on fossil fuels.

Tourism gives us an ideal opportunity to try out 'new' ways of transport; when we travel as tourists we have time and the opportunity to try the different things that everyday life makes difficult in the midst of its routine and rush. The essential framework we need for this change is public transport, provided with adequate frequency on a networked basis, with the possibility of moving from one mode of transport to another, linking buses or trams and trains, for example, by time and place. The National Trust has taken the vanguard in deliberately encouraging public transport for a higher proportion of visits to its properties. The annual *Handbook* of the Trust gives the necessary, detailed information to facilitate these visits. A further necessity in linking in public transport is sufficient space and easy access to carry personal transport mechanisms, such as bicycles and cross-country skis, which can be so awkward to get from one place to another.

The challenge and style of this effort is appropriate for our 'adventurer' role, already discussed as a dimension of contemporary tourism and society. The creativity needed for such travel also has a resonance with our emerging personalities. After all, who could argue against the presence of the element of personal choice in this style of travel? In a society where we are conditioned to prize range and diversity, a prime benefit of travel is the variety we experience in the different elements that make up our journeys. Moreover, cycling not only gets the traveller from one place to another, it also delivers the sensual appeal of fresh air, and the scents that rise from the countryside as we wheel by, an experience that is denied to us in fast car travel.

Within cities and their surrounding urban areas, the argument against travelling in private cars or, at least, the argument for large-scale reduction in their use, has largely been won. Not that this 'victory' is universal, since there are contrary developments such as the growth of counter-attractions to the city (out-of-town shopping malls, for example) that depend on their customers arriving by car. Severe solutions to traffic problems have been suggested, such as closing cities to private cars altogether or putting tolls on their use in inner-city streets, but pressure from such quarters as city centre shops and businesses may stop these solutions from being much imposed. Some pilot schemes in the use of traffic tolls are already being tried. Here again, the eventual outcome will depend on the demands of society and consumers.

Chapter 1 posited that one of the main attractions of the city is the opportunities it gives for watching and meeting people. This activity, of course, is an overall key function of Dynamic Tourism. Again and again, we have seen that those parts of cities that have been closed to traffic go through a rebirth; they flower into new energy and are refashioned. In these places, walking becomes the mode of transport. Experiences associated with walking, and pastimes that are enhanced by the pace of walking, become the preferred activities. People seem to enjoy such experiences, and increasing numbers of people want to see them as fundamental to their tourist activity.

In heavily visited parts of the countryside, vehicles driven by tourists may be visual drawbacks, an encumbrance and particular impediment to a quality experience for the tourist, and even more so for the permanent resident. Methods other than bringing public transport access to entry points may be needed; it may be necessary to encourage walking or to provide mini-bus transits. In the United Kingdom's Peak District National Park, which is the world's busiest national park in terms of visitor days, the concept of applying a congestion charge is under consideration (Ward, 1995: 9; Bevan, 1999: 9)

However, solving one problem by encouraging people to walk rather than drive may create another problem in fragile or heavily visited areas, such as the Peak District National Park. A solution may be offered in one approach to walking and public access, used in Sweden under the name of *allemansrätt* – essentially *every person's right* to roam anywhere (except for certain situations and circumstances, such as in deference to farming requirements). This approach fits Dynamic Tourism's idea of including all parts of the world in the tourism equation, so that the load is spread more widely, and also, of course, to offer maximum versatility and difference.

Notwithstanding the pressures it creates in certain very popular areas, walking is a key activity among the recommendations of Dynamic Tourism. Why? It is enjoyable in itself. It increases the potential for unexpected and illuminating encounters. It is also good for health, and thus combines several dimensions in developing toward 'wholeness'. This desire to walk is yet another concept that already seems popular among tourists. English Tourist Board (ETB) research showed that walking is the most popular holiday pursuit for the British, with £375 million spent on 3.2 million walking holidays taken in the UK in 1996. Significantly too, the research showed that, 'In addition 15 % of all domestic holidays involve some form of active walking' (ETB, 1998: 2). In discussing walking, we should not forget to include the 'ramble', a good example of the serendipity that features in Dynamic Tourism. A ramble enjoys the slight connotation of a less purposeful movement than a walk and may perhaps be embarked upon as an experience of gentle relaxation, an invitation to move aimlessly, and welcome whatever meets our interest.

Extent and flexibility

Now is the moment to turn to consideration of other specific signs of Dynamic Tourism in today's travel industry, and to do so at a major level. More than a century ago, it was recognised that the whole world, in its entire range, constitutes the tourism product. Turner and Ash (1975: 50) report that the early tour operator Thomas Cook announced 'God's Earth in all its fullness and beauty is for the people'. This statement also voices a perception of tourism as a democratic medium, not one confined to select groups. There is the hint, too, of an avant-garde inflection, the suggestion that tourists should not be instructed and patronised.

As was suggested in Chapter 1, the critical acceptance, implicit in the idea that tourism covers the globe, is that every location is potentially a destination and should be recognised as such. This may be a difficult idea, too novel for some people to grasp in the countries that traditionally generate tourism. Sudjic opines (1995: 2), 'the next cultural clash over tourism isn't going to be on the beaches of Asia or the Costas. It's going to be back in northern Europe, where it all sprang from in the first place'. He makes the telling remark, 'Wait and see what happens to Britain when the Japanese and Malaysians start treating Wiltshire in the way Peter Mayle treats Provence'.

It is clear that a part of the flexibility, change and accommodation of the entire world in tourism, which is the essence of Dynamic Tourism and is necessary to it, will call for all the generators and receivers of tourism to reinterpret their view of themselves. They may want, or need, to make changes in order to serve in a new role opposed to their former one, or they may want to act simultaneously, both as generator and receiver. Flexibility is already on the agenda. At a World Tourism Organization (WTO) General Assembly round table of experts to consider trends in tourism, Bruce Cameron identified three critical elements that influenced travel in the 1990s. One of these elements was 'More flexibility in travel plans, more independent options available' (WTO, 1995: 3).

Peace and tolerance: meeting and balance

The linking of peace to tourism is a somewhat controversial one. As the preceding paragraphs suggest, tourism encounters may potentially become trying and offensive. The tensions of the wider world are much stronger, and more difficult. However, Dynamic Tourism, with its emphasis on encountering other realities, and the rewarding benefits that result from these encounters, encompasses the categorical belief that tourism, in providing us with opportunities to see and thus understand our global neighbours, can serve the aims of preserving and/or re-establishing peace.

Both the 1994 and 1999 Global Conferences of the International Institute for Peace Through Tourism outlined possibilities in the arena of peace and tourism. At the 1994 Conference, the tone was set with an opening Interfaith Service, 'to recognise and celebrate the diversity of cultural and religious traditions which are part of the human experience on our planet'. Presenters delivered individual statements of welcome to this service in Mohawk, Christian, Baha'i, Muslim, Jewish, Buddhist, Hindu and Sikh. We know that our diversity contains the seeds of conflict. Although tourism brings that diversity into sharper focus, by bringing us up close to difference, at least it allows us the chance to know and understand at close range. The other option is to allow distance and division, permitting misconceptions to flourish and fester. Tolerance is connected to peace, and therefore, the establishment of the Los Angeles Museum of Tolerance has been a significant development.

Our recognition and acceptance of others as equals in tourism must emerge and be recognised. This is a mandate of our society's multiculturalism, but it is also a necessity because we must still address an imbalance remaining from the past, when a colonial visitation of supposed 'lesser'

*Telling visitors what they need to know: Shenandoah
National Park, Virginia, United States of America*

orders and natives was a frequent characteristic of tourism. One indication
that the imbalances caused by colonial history are being redressed is that
the World Heritage Site of Uluru (Ayers Rock), a critical site in Australian
Aboriginal belief, is now predominantly managed by Aboriginals.

It is a generally accepted proposition that tourism is continuing to grow,
and attention is attracted to its commercial and socio-economic potential.
The Commission of the European Communities has seen reasons for continuity of this increase and identification:

> Conditions are favourable for further growth in the numbers of tourists,
> particularly at international level, because of the ageing populations in
> industrialized countries, higher levels of education, more widespread
> paid leave and shorter working hours. These are all grounds for fore-
> casting a further expansion in short-stay holidays, more frequent holi-
> days, and greater access of tourism to young people and retired persons.
> (Commission of the European Communities, 1995: 5)

Protecting, conserving and sustaining

As we know, the impact of an increase in tourism can be harmful as well as
beneficial. Tourists need information to help them play their part in reducing
or avoiding adverse effects. This means that information presented at tourist
sites and vulnerable places must not only mirror our environmental
concerns by explaining how our activities and way of life threaten these
places, but must also show the possibility of taking these concerns a step
further in the form of encouragement to actually reduce the hazard. The
Shenandoah National Park in the United States of America uses shock
tactics. An information board at a scenic overlook there explains why the

view of the landscape is impaired: tourists cannot often see very far into the distance because of modern air pollution caused by human activity.

It takes mature judgement on the part of communities and the heritage profession to protect sites from tourist impact. One example of such maturity is seen in the measures taken upon the accidental discovery of prehistoric wall paintings in the Grotte Chauvet at Vallon d'Arc in the Ardèche region of France. It was quickly decided that to protect these paintings from damage by human breath and presence, they should not be opened to public visits. This example may be compared with the fate of the prehistoric cave paintings in Lascaux. After many years of visits to the cave, the damaged paintings were finally put off the tourist circuit in the 1970s, and a replica provided nearby to satisfy public interest.

Special Aspects and Instances

Change in continuous motion

The examples above display change, or efforts to bring change about, and so they mirror the essence of Dynamic Tourism. When Neil Cossons was the Director of the Science Museum in London, he openly recognised the need to change, in his Introduction to the Science Museum guidebook, ' ... because science is about change, the Museum tries to reflect current developments by constantly adapting its displays and introducing new ones' (Cossons, 1994: 2). The particular interest of this remark lies not so much in the fact that the presentations of the Science Museum were changing, since most museums adopt the practice to some degree, but in the way it reveals an acceptance of repeated change, and an understanding of its necessity. Dynamic Tourism is not just about change, which may be episodic, but about ongoing, continuous change, as a response to changing audiences, public demands, and circumstances.

Airports

Moving on to another, associated dimension of Dynamic Tourism, we can assert that airports have certainly changed. They have been among the quickest and most sensitive elements within the tourist industry to reflect our contemporary needs and dispositions. Airports now mean more than an application of the practical concerns of travel; they are now understood to be a positive part of the holiday experience, and capitalisation is backing up this premise. We spend money at airports, going outbound, because we are already in a mood to enjoy ourselves as we reach the airport, or because we notice that we still need some critical items for our holiday. On our return trip, we can be encouraged to purchase souvenirs that flaunt the fact that we have been on a trip, or as gifts for other people, perhaps to assuage our guilt at having travelled when they did not. Retail and food outlets now

To walk and cycle, self-guided and informed: St Margaret South Elmham, Suffolk, England

abound in airports, ready to meet the needs and impulsive whims of travellers. There are outlets of the famous Harrods store, which is a tourist venue wherever its location, at both the Heathrow and Gatwick London airports. In tune with Dynamic Tourism's identified tendencies, less-mainstream retail outlets have emerged at air terminals. The Nature Company, for example, was present for a while at both Heathrow and Gatwick airports, selling products related to the environment and holism.

Bicycling

Transportation was discussed above as one indication of societal transformation, along with the attendant suggestions of changes, both in the choices of transportation available to us and in our general attitudes to it. One example of our new perceptions in this area is the existence and work in the United Kingdom of the charity Sustrans. Sustrans aims to provide 'paths for people' and to 'build where the need is greatest'. The charity provides paths for walkers and cyclists that are also accessible to the disabled, and meant to be encouraging to nature. The Sustrans endeavour for the millennial year 2000 was to deliver a national network of traffic-free cycle routes amounting to 5000 miles, which will be extended by another 4000–5000 miles in five years, and then 50% of people in the UK will be no more than two miles distant from access to it. As part of its general progress, Sustrans established the C2C, a round-trip cycle route, from coast to coast in northern England. A corollary of the Sustrans idea, and part of their general practice, is the placement of sculpture along the C2C's wayside.

The C2C is in tune with Dynamic Tourism by reason of its simple, individual and relatively environmentally-friendly philosophy. Part of this is because it is designed to be used by bicycles, but it is also true because the C2C brings tourists into new areas, some of them aesthetically unappealing. It is thus an example of the extension of the portfolio of possible tourist destinations. From the point of view of the provider and host, this example presents opportunities to do business. Similarly, it provides chances for social enhancement and economic development in places that have been neglected in the wake of post-industrialisation.

Simplicity and abstraction; fluidity, provisionality; the garden and nature

Yet another feature that draws particular attention is the garden, due to its perennial appeal and particular attraction for contemporary society. We have already briefly considered the overall features and components of gardens and analysed the character of their appeal. One special dimension, which is perhaps yet more significant, since it is connected to other matters that we have discussed concerning contemporary society and the principles of Dynamic Tourism, is the unexpected prevalence of Japanese gardens, and among them, of Zen gardens.

Zen gardens are distinguished by their particular dimensions of simplicity of composition, associations of rigour, auras of spirituality, direct connection to a formal religion and philosophy, and their uses as mechanisms of meditation. They occur in quite surprising places, in relation to tourism, and are found far from Japan. For instance, there is a Zen garden in the new National Botanic Garden for Wales, and a Japanese garden is among the features at the Disneyland Hotel in California. Zen styling appears in countless places as a landscaping tool; in the United Kingdom it may be seen in a range of sites, from Exchange Square in London to a McDonalds car park in the East Midlands. The Ryoanji Zen garden at Kyoto in Japan is a major crowd-puller as a destination in itself.

The key aspect to Japanese gardens is that their features, compositions and arrangements have meanings. These meanings are connected to the philosophical and spiritual side of the world, and so signify philosophical and spiritual depth, adding special interest to the gardens. In their calm, beauty, natural qualities and narrative symbolism they are antidotes to the technology and speed of everyday life. Japanese gardens offer contrast and completeness, both of which are needs that Dynamic Tourism has identified among those it strives to serve.

The Kyoto Garden, installed in London's Holland Park, is a *stroll garden*. It gives overt attention to another dimension of existence, that of time, and so leads thought back to the essence of all gardens; they never stand still. This too is reminiscent of those characteristics we have identified in our definition of Dynamic Tourism, the need to present the aspects of provisionality and constant change. There is a plaque in the Kyoto Garden that says, 'Time is an element that adds to the garden's natural beauty. The climate affects the original design over the years, making the garden's landscape even more natural and fitted to its locality'.

Since Zen gardens have gained appeal, even though their features are non-figurative, they show that visitors have the sophistication to cope with and to explore abstract meanings. These gardens indicate that we do not necessarily any more need figurative representation and information, at least not in every instance. This point is important, and very relevant when considering how to choose the style of presentation for tourist attractions.

Two exhibits of the 1990s demonstrated just how communicative abstract style can be, and also how conducive to the experiential it can be. The first example was the final portion of the Victoria & Albert Museum's popular *Japan Today* exhibition of 1991–92, 'Dreams' by Toyo Ito and Associates. This placed the visitor in a space inhabited by nothing but sounds and projected images in which they were 'bathed in a shower of information and white noise' (Victoria & Albert Museum, 1991). 'Dreams' was an overall spare presentation, leaning strongly towards abstraction, and it was a startling contrast to 'Chaos', which was a depiction of modern Tokyo offered to visitors as the middle portion of the exhibition. 'Chaos' presented visitors with elements such as the sensation of massage chairs and an onslaught by *manga* (illustrated comics intended for adults as well as children). This middle section, again, contrasted with the first display, 'Cosmos', which evoked traditional Japan through a few representative artefacts and features that included a modern reconstruction of a seventeenth century teahouse.

The other exhibition, *Room on the Run* at the Swedish Museum of Architecture in Stockholm in 1998, addressed the subject of rooms and space and what they represent. Visitors were invited to walk around the ultimate abstraction and simplification of shelter and space, as represented by the following: the area covered by an umbrella, the temporarily chalked out squares of a game of hopscotch, simple forms of shelter, a Modern Movement dwelling displaying inside-out inter-relationship and abstract style, and the supreme simplicity of an enclosed space with no furniture other than the people who entered it. The Victoria & Albert Museum and Stockholm exhibitions were directional, in that they illustrated notions in such an abstract and simple way, while also portraying provisionality.

Elemental: prehistoric and modern

In today's society there seems to be a profound interest in simplicity and abstraction, the basis of a trend that is now emerging into prominence as an antidote to an oversupply of detail and complexity. Exhibitions such as those at the Victoria & Albert and the Swedish Museum of Architecture must both reflect the existing interests of their time, and also serve to strengthen and widen those interests. The sheer strength and purity of a vision grounded in the elemental seems to have a basic appeal to us. We are drawn, for example, to caves and standing stones. Zen gardens touch the appeal with simple rocks in their design. Stonehenge, with its ring of megaliths, has always exercised a massive fascination and appeal to tourism. Setting aside the association with Aboriginal culture and religion, might not these same characteristics of abstraction and bare stone be part of what draws us to visit Uluru? It is essentially a natural feature. Nature can endlessly offer for our admiration the same dimensions of appeal of elemental features lacking description. This appeal is in the shapes of hills, accumulations of clouds in the sky, rock formations, and the particular interactions of the natural elements of earth, sand and sea in the landscape offer items that render more places as destinations in the tourism portfolio.

From this point of view, it is helpful to remember that there does indeed seem to be a basic attraction in natural, unattended features in the landscape, an appeal in elemental or non-figurative forms. Rural and 'untouched' areas, which for reasons of economic and social development are often among those places that are most eager to attract tourists, have a raw element of appeal that lies precisely in their unmanicured, raw state. This overall interest generated by the abstract and primitive draws our attention, and leads inevitably to the suggestion that both prehistory and the modern (by this we mean Modern Movement artefacts themselves, and later items that bear the imprint of that ethos) have considerable appeal – far more than is generally believed. We can put together with these elements others that are characterised by simplicity and abstraction, such as natural features in their many forms. These categories could provide the basis of extra attractions; they are logically part of the diversification recommended by Dynamic Tourism. We will return to the theme of attention to the modern later, but one preliminary demonstration of the appeal of modernity would be the success that modern monuments like the Sydney Opera House (a World Heritage site in Australia) and the Pompidou Centre in Paris have had in attracting visitors.

Temporary by design

We should not be surprised that, among the examples of tourist attractions that display the dynamic characteristics of fluidity, change, provisionality and simplicity, are many that are temporary anyway. During the re-development of the devastated Potsdamer Platz area of Berlin, for example, tents were set up on the cleared site. These temporary shelters held exhibits that showed the area's former interest and importance.

The title of 'European Capital of Culture' is a designation that lasts for only a year. Among those cities that have enjoyed this honour, Copenhagen and Stockholm used the opportunity to deliver dynamic elements and to try out innovations and experiments for the future. One example of this was the centrally located visitor presentation about old Stockholm. It was fresh in its simplicity, housed in a wooden cabin with benches for seats and, with a twin-slide presentation of scenes as the exhibit. Below the slide show were maps showing key locations, with appropriate music added to enhance the experience.

Both the Berlin and Stockholm examples stand in contrast to so many weighty temporary exhibitions that are mounted at great cost and so usually require corporate financial sponsorship. It is especially pertinent to point out that effective, communicative, high quality exhibitions do not necessarily have to rely on heaviness, complexity and great expense. This is important information for tourism providers who are hampered by limited resources, as is often the case in small, rural communities.

To the Fore

Frequently it seems to be the case that tourism-providing countries and destinations which lie out of the main stream or which are more recent in developing tourism as a resource are those that reveal an appreciation of new ways and products. These are the places that demonstrate Dynamic Tourism in action. Maybe these sorts of tourism providers realise that they cannot compete in the 'old' way and on traditional grounds, owing to resource or infrastructure deficiencies when measured by standard criteria.

Nature is one resource to be used in these cases. Poland, for example, positions itself as a nature tourism destination, the best in Europe, 'where you can admire wild nature and human inhabitants livings in harmony' (Polish Tourism Information Centre leaflet). Iceland, which seeks an antici- pated number of visitors equivalent to its total number of inhabitants, promotes itself as 'a rare attraction, ... [an] unpolluted environment, natu- rally endowed with pure water and foodstuffs and vast resources of envi- ronmentally friendly energy' (Icelandic Tourist Board release: November 1995). The Finnish Tourist Board promotes Finland as, 'the world as *nature* intended' (Finnish Tourist Board advertisement).

Spirituality is another factor that may be exploited in this kind of tourist destination. In Tibet, where the Buddhist religion is a key cultural element, tourist development is concentrated on cultural monuments, with the capital Lhasa as the focus, along with trekking and the natural environment (WTO, 1994: 156–8). Interestingly, Tibet is using 'tented camps' (ibid. p. 161) to increase available accommodation at its peak tourist times. The dimension of spirituality and the particular interest in Buddhism, which has become prev- alent in developed countries that generate tourism, can also be fed by travelling to Pakistan, 'Land Of Culture And Adventure' where there is a 'Buddhist Study Tour' offered by PIA (PIA travel brochure).

In Australia, a distinctive style of tourism has developed alongside the standard Western tourism template. Maybe the geographical size of this continent and its distance from the United Kingdom, which gave it its modern-day cultural foundations, have led to this development. Perhaps this distinctive style is also due to the 'walk-about' culture of the first Aboriginal inhabitants. There is no mistaking the general independence of outlook that Australia has generated by necessity, along with the 'machismo' of the Outback where the inhospitable nature of the land has demanded tough self-reliance from both its residents and its travellers.

Backpacking is a widespread element of travel over much of the globe, but it has its apogee in Australia, where it is a particular, integral part of travel. In Australia backpacking has developed its own special, distinguishable culture. The very character of backpacking emphasises independence, self-reliance, flexibility of choice and movement, adventure, contact with nature even in its extreme manifestations. Backpackers seek chance encounters on the journey and want to learn from them; they patronise simple accommodation, engage in sport, make music and share it along their way as they forge new friendships with fellow wayfarers. The essence of this style of travel is serendipity and pleasure in the unexpected, in large part provided by the complexity of travelling around such a large and physically demanding territory. Likely products of such a journey are increased self-discovery and self-development.

As we can easily see in this characterisation of Australian-style backpacking, many features of Dynamic Tourism are already present. It is clearly an experience. Another general aspect of this activity is that it occurs, and is successful, in a habitat that is not traditionally promoted for tourists. The climate is not comfortable, there are no great concentrations of man-made monuments and buildings, nor are there many other visible cultural attractions. Once again this example demonstrates that places devoid of routine appeal as a destination for tourism, with no sizeable tourism infrastructure in place, can still have the capacity to be tourism venues.

Basic forms of transportation other than walking are showing signs of reviving interest, notwithstanding one irony inherent in modern society–the fact that each long journey to the holiday venue requires the use of motorised transportation. In any given summer season, how many bicycles ride along the highways and byways of Europe strapped to holiday cars and camper vans? Their great numbers attest to the bicycle's attraction as a means of transportation, once the vacation destination is reached. Indeed, in the Netherlands, biking holidays are a local speciality, since the flat landscape lends itself so well to cycling; but the Dutch are not the only ones to offer this particular treat. The adventure holiday company, World Expeditions, whose name and slogan 'Because life is not a dress rehearsal' show a new tourism approach, offers both walking and cycling trips. Some of these are to the relatively unpenetrated tourist destinations of Vietnam and China. Such trips are not for the multitude. However, as a sign that cycling tours, even to less common destinations, are indeed entering the industry mainstream, we note that the major tour operator Thomson has offered a 'Danube Cycle Tour' as part of its Lakes and Mountains programme.

Altered Attitudes, Approaches and Styles

Yes, the mainstream tourist industry is absorbing contemporary society's changing tendencies and sensibilities. Another indication of this process is the manner in which 'old' tourist destinations and mass tourism sites are now being promoted and presented. This may be seen in Spanish Tourist Office advertisements that tout 'alternative', unconstrained, light attractions, the chance to return to basics, to experience sensual, environmental dimensions. The main caption is the offer 'You can go ... wild', and above it is the mention of 'far from the madding crowds'. The supporting text reads:

> A country with the timeless appeal of Spain deserves to be viewed at a pace far removed from the pressures of everyday life. To this end, more and more visitors are getting back to nature and travelling light. There is nothing to touch it for really feeling the warmth of the landscape and the people who live in it. And, ecologically sound considerations apart, you'll find it decidedly easier to park. Be it in the shade of a tree, Romanesque church or village bar. (Spanish Tourist Office advertisement)

Another firm, Vacanze in Italia, adopts a similar tone, stressing the spiritual aspect of its destinations in the text of its advertisements:

> A hilltop café serves Chianti by the glass, soft voices float up from nearby tables. As the sun sets, swallows soar around the campanile. Italy/Tuscany is a refuge for the spirit. (Vacanze in Italia advertisement)

This last example introduces yet another aspect of today's tourism, the specific mention of wine and, by implication, of the food that accompanies it. These pleasures of the table are a burgeoning dimension in society. Tourism not only reflects this contemporary interest, but also has strongly encouraged and developed it. Our fascination at home with types of foreign food and wine that we have consumed abroad already forms a dynamic. When people are off again on holiday, they are already predisposed and stimulated to try still more new sensations. Perhaps the attraction of our home consumption of exotic food and drink is that these provide us with a quick, easily accessible sensual connection to our experience of abroad, and so we are allowed some faint repetition of an event – the holiday – that we regard as important. Food stuffs and wine are easily transportable from foreign lands. Once in our homes, they serve as a means to transport us in imagination to our vacation and its environment.

Wine holidays probably derive part of their attraction from the fact that they represent a sensual experience. They provide an opportunity for us to prolong the holiday experience by purchasing wine and bringing it home with us. Many rural areas have already realised that local produce is a solid and appealing item as a tourist attraction, suitable for places that may have few distinctive features of interest to attract visitors. Dynamic Tourism promotes general diversification of appeal, and the particular food and drink of a given area can act as a powerful tourist attraction, while also serving as a means of distinction, differentiating one chosen place from other seeming equivalents.

The majority of visitors to the Bordeaux region or the Côte d'Or in Burgundy would probably not be there for the landscape alone. In neither instance does the landscape itself, with the exception of the chateaux, have a particularly strong appeal. The travellers are there because of the wine and associated culture of the area. If we need global confirmation of the crucial importance of wine as a key element in Bordeaux, we may note that the Saint-Émilion area of Bordeaux has been designated a World Heritage Site, a cultural landscape embodying a long tradition of grape cultivation and wine production.

Just one more advertisement provides a final example of the themes that we have been exploring: the Greek National Tourist Organization, in conjunction with Olympic Airways, uses two pictures to form its image. In one of these, we see a traveller, accompanied by luggage, including a back-pack. The leading slogan applied is 'Not just a tourist. An adventurer'. In the continuation text is the exhortation to 'Experience Greece'.

Impediments and Wrong Turnings

Here, then, is an overall picture of tourists, as they reflect their wider dispositions as members of society, wanting those things that Dynamic Tourism suggests. The illustrations shown here reveal some recognition of the type of attitude that is necessary within the tourist industry and in the world at large. These examples demonstrate an initial response by some tourism providers to the changes in their audience. It would be easy, there-fore, to imagine a smooth path of progress, to expect a fuller recognition of the forces at work, and thus to foresee necessary and suitable changes occurring as a reasonable consequence of all that has gone before. However, it can be argued that there have been clear indications of a need for change for quite some time, while as yet only an inadequate response has been made. In 1964 Lewis Mumford remarked:

> The new task of the landscape architect is to articulate the whole land-scape so that every part of it may serve for recreation. ... One must think of continuous strips of public land weaving through the whole land-scape and making it usually accessible to both nearby residents and to holiday visitors. ... There is the beginning of this new process of using the whole landscape as a recreational facility in the layout of bicycle paths in the Netherlands; and there remains in certain parts of England ... a system of public footpaths over hill and dale, through field and wood. (Mumford, 1964: 169–70)

Note how Mumford's comments draw attention to two aspects. The first aspect is that the impulse toward bicycling, which he identified as neces-sary and already present, has since stalled. It has even gone backwards to some extent in England. In the Netherlands, in contrast to England, the emphasis on bicycle transport has continued to develop, still including the tourist market in the increase. The second aspect relates to the prospect of public access. In the United Kingdom, England especially, where the land area is small and the population relatively large, matters of land ownership and public access to it are still being argued. In contrast to the United Kingdom, Sweden has a larger landmass and smaller population and, as we have already seen, *allemansrätt* permits public access anywhere in the countryside, with certain exceptions, such as to avoid interference with farming and cultivation.

One difficulty in getting acceptance for the proposition that change is useful for tourism may be rooted in cases where a relevant idea is adopted, but deployed only superficially and then subsequently ruined in its use and development as a commodity. The foregoing discussion suggested, for example, how aspects of deep spirituality can be trivialised and spoiled through over-attention, increased popularity and altered use. Once again, we can find examples in the present fashion for Buddhism and Zen. The depredation that tourism can wreak in this connection is skewered in remarks made by one of the characters in Douglas Coupland's *Microserfs*:

> Thailand? I love Thailand! I'm dying to build a chain of resorts all over Thailand and Bali, kind of like Club Meds but a little more nineties. I'm gonna call them 'Club Zens,' right? 'Cause of the Buddhist thing. There's all kinds of statues and monuments over there I could use to make it look authentic – like you're in a monastery, but with booze and bikinis. Now that's nirvana! (Coupland, 1995: 216)

Experience, and More and Different Participation

Notwithstanding Coupland's lampoon, it does seem that we are seeking more depth to our tourism, whatever form it may take. We are no longer so much sight-*seeing* as sight-*involving*. We want to know, feel and understand what we visit and perceive. We want to measure the significance of these perceptions to our own lives. This reflects our search for experience, which may perhaps be characterised as the key tourist desire to be met by Dynamic Tourism. Among important attractions for contemporary tourists are the emerging sites that deal with one of the last century's most critical events (and one of the most difficult to address) the Holocaust. Interestingly, the Jewish Museum in Berlin opened its building of 'telling' space some time before the interior exhibition was ready for installation. The United States Holocaust Museum in Washington DC has been a major world exhibition since its opening in the early 1990s. Its style of presentation works deliberately to engage emotions, to elicit shock from its visitors, and to invite their identification with the subjects of its exhibits. It is meant, in other words, to be experiential.

Many travellers now expect and require a more intimate experience of other countries and other cultures as part of the holiday experience, as opposed to staying in a confined holiday precinct or area. This greater intrusiveness of tourism can carry with it a greater impact on environments and communities, perhaps resulting in an increase in harm to them. Many types of working holidays are now offered, with the purpose of offering amelioration and antidotes to this problem, and which, again, show propensity for involvement. For example, there are the National Trust Working Holidays where holidaymakers busy themselves with restoration of heritage properties and conservation of the environment. For children, there are the Acorn Camps, where the core holiday activity is to deliver care and help to National Trust properties.

It is interesting to note that there is an overall number of 38,000 National Trust volunteers (Drury, 1998: 19), a fact that signifies the level of commitment generated by involvement with the Trust. It is reasonable to suppose that many of these volunteers first became involved with the National Trust as tourists. Many of the Trust's large number of volunteers are retired or Third Age members of society (National Trust, 1998). This fact points to the characteristic tendency of senior citizens to participate strongly in society and thus to wield greater social influence. Traditionally, it has been the young in society who tend to blaze new trails. Today's young people still have huge influence, but now they must share it with people over fifty.

As we noted in Chapter1, today's 'greys' are not like the elderly of the past. Among the present generation of senior citizens are the last century's post-Second World War Baby Boomers. They were in the forefront of the revolutions of the 1960s that culminated in the rebellions of 1968. So it is not only the young, but also the new elders, the *soixante-huitième* brigade, who exert a critical influence in the creation of the altered mind set of today's tourism, characterised by changed interests, strong independence and greater adventurousness. Do not forget that the elders of the Westernised world are likely to share with the young the characteristics of freedom from commitments; they too are 'foot loose and fancy free'. The elders have ridden the crest of the wave of the post-War welfare philosophy right through life, and they are benefiting in terms of financial, educational and physical well being. In their changing conditions and expectations, these elderly parallel the young, but with the important distinction and attraction of probable affluence. Therefore, these 'greys' are an attractive market for tourism providers to pursue; it will be profitable to understand their concerns.

Key Changes and Changers: Why?

Through the use of a number of illustrative examples, this chapter has attempted to clarify the ideas and directions of Dynamic Tourism. The objective has been to identify those changes that Dynamic Tourism is designed to serve, and to consider some particular aspects of tourism that indicate that necessary alterations are already underway. When a particular disposition for change, or a ready appreciation of it, is present in a country or geographical area, there is an aptitude for the recommended response to Dynamic Tourism. Places in the frame of mind for change, which are indeed innovating and altering their ways are: Australia, Canada, The Netherlands, Japan and the Scandinavian nations. It is not the purpose at this point to look for reasons, other than to notice the following features among the listed countries: a characteristic distinguishing physical feature, a degree of geographical distance or isolation from outside influences, and an impulse to differentiate and be identifiable in contrast to a strong, high-profile neighbour.

It is possible that the situations outlined here have developed strong, individual cultures in these countries and thus, by necessity, attitudes of creative and innovative thought. Both Australia and Canada are multicultural countries, originally amalgams and accommodations between Aboriginal peoples and colonial newcomers, but now much more complex mixtures. In Scandinavia, we see the Sami people as first occupants, joined later by other peoples, thus producing a cultural mix in much of that territory as well. All of these countries, in fact, are currently cultural meeting places. Their core identity is undergoing cultural fusion or, at least, complex interaction.

One notable characteristic of all of them is that, for one reason or another, they were encouraged to think for themselves, and have become accustomed to doing so. Perhaps this habit of independent thought has equipped them to take the lead in noticing signs of the need for change, and in acting on those signs. These are, on the whole, affluent lands, where people are generally secure in their basic needs, and thus not preoccupied with necessities. So it is also possible that they have enough surplus to enjoy the luxury of developing sophistication, of noticing causes for change, and of thinking out innovations.

Chapter 3

Knowing

Suppliers and Tourist: Duties, Connections, and Separate Obligations

Up-dating, all the while

Knowledge fuels the shift to Dynamic Tourism, and it always will. We will no longer be able to coast, if we wish to practice a type of tourism that is characterised by change and adaptation to fluid tastes and circumstances. In the past, it was possible to depend on a familiar system, fundamentally static in its style and rituals, and thus transformed by routine into second nature for tourism's regular participants, but this will no longer be the case. Fresh knowledge must be continually available, and we must continually absorb it. As providers and consumers, we need to know in order to respond to change.

The necessary knowledge is two-fold. The industry must know about the tourist, and the tourist must know the products of the industry. Today's available information on both sides is not adequate, particularly as regards tourist tastes, patterns and inclinations. There is general understanding among the industry and its partners of their particular difficulties. For example, *Tomorrow's Tourism* (the UK Government's tourism strategy publication) states plainly that the tourism industry, 'suffers from lack of market information, or the ability to interpret and respond to it' (DCMS, 1999: 9).

One perceived problem is the lack of adequate mechanisms in the industry to monitor the outcomes of initiatives. How are new projects to be judged? What impact do they have, be it positive or negative? How may their success be measured? Providing answers to these questions, in the complex situation of contemporary tourism, is not easy. Financial resources adequate to implement systems are not enough. Research designs must also be put in place, so that a statistically significant body of reliable data is obtained on a balanced basis and in a meaningful form. On the one hand, the researcher faces the practical difficulties of gathering data in as balanced a way as possible. On the other, the old adage of 'rubbish in – rubbish out' holds a central position. Unless the questions we pose are formulated with an adequate understanding of tourists and their circumstances, the information solicited may arrive badly skewed or with key elements missing.

Thus the industry faces a Catch 22 situation. In order to obtain accurate information about tourists, researchers already need to have a great sensitivity to and understanding of tourists. At the beginning of this book, we recognised the necessity in tourism of a greater degree of revealing and catering to general consumer tastes. The body of information already available on our general inclinations as consumers in other areas of the economy can provide a useful starting point for similar studies of the specific issues of tourism. Such information can help us know what we would like to do as tourism buyers, if suitable products or opportunities were to be made available. For example, information about catering to business trips and holidays, two essential sectors of tourism, may be couched as a comparison of how we allocate funds on behalf of our employer, in contrast to spending our own money for leisure and as a discretionary option.

It is critical for the industry to know where tourists are likely to seek pro-actively for information about the industry's products. Still more crucial is an awareness of what media and data sources are in use, even when we are not overtly looking for tourism information. In this latter situation, the tourism industry objective is to attract unpremeditated purchases or 'impulse buys'. The industry's essential goal is to create a climate of information and promotion, which lures us to spend money on tourism products. Another goal may be to divert consumer attention from a rival option. The discretionary tourism purchase may be made in opposition to another item that might have filled the same need in the buyer, or even a rival need. Beyond the sales resistance of the customers, there is the additional challenge within the tourist industry of the attractive offers of competitors. These rivals will always be honing their expertise and vying to meet consumer desires more closely and with greater accuracy than their competition.

Generally speaking, therefore, the tourism industry needs to know our tourism wants, and also where we will seek the information to fulfil them. The essence of Dynamic Tourism is to pay attention to our tastes and capacities for change. From the industry viewpoint, it may seem difficult to lock into and interpret this fluid situation. It is a recognised provider demand. Out of necessity, the advertising industry seems especially adept at acting in support of providers to meet this kind of challenge. This media sector is a byword for staying tuned into society, as well as for sophisticated, intelligent anticipation of consumer facet niches. This is helpful in the task of 'reading' the consumer and obtaining the necessary depth of knowledge to successfully target individual groups. Although the *expressions* of our tastes vary and change, their *core motivations* remain largely the same. Dynamic Tourism caters to those well-known needs of holidaymakers, their wish for status, self-fulfilment, excitement and stimulation, or relaxation. When applied to business travel, with its characteristic key need for travellers to show status, Dynamic Tourism serves the sector's needs for flexibility, adaptability and efficiency, while allowing creativity in making optimal choices.

Responsibilities, and imbalances

In 1995 Ken Carey predicted in *The Third Millennium: Living in the Posthistoric World*:

> In the dawning light of this millennium, the framework surrounding human affairs can be seen as a chessboard of sorts, each global interest represented by a piece on the board. Each piece, each player, as this new era dawns, is possessed with access to unlimited information. And each knows that using that information intelligently is the key to success. Institutionally invested human interests have at their disposal not merely computer technology to help them organise their information more accurately but also a rising flood of information itself, unprecedented in the history of nations. (Carey, 1995: 100)

Carey's comments stress several key points beginning with the huge amount of information, and the emergence of an increased personal and organisational responsibility for effective use of that information. The second point is to suggest a general necessity of maturity in our use of information. There is indeed a burgeoning quantity of data. Coupled with the inter-related interests of the players on Carey's chessboard, this 'rising flood of information' raises a potential problem. Even in an information-laden, computer-linked globe, the same amount of data is not necessarily equally available to all. The corresponding difficulty lies in removing imbalances in the ability to handle and interpret data. Even the modern, successful tourism industry is not yet adequate in its capacity to gather necessary data, let alone to understand and use this information optimally. Another dimension, already mentioned, is the requirement that information be of usable quality. Tourists, the tourism industry and its companion interests must set themselves on a higher level of rigour in regard to information.

The industry personality and constitution

The tourism industry and its participants must deliver appropriate and adaptable products, which society as a whole, in a given time period, will regard as acceptable. To do this, both a fine data bank of information and a sophisticated ability to use it must be present. An infinite number of statistics may be gathered but, unless they are 'seen' and recognised by an imaginative mind, their story may not be appreciated, and their implications may be lost. Remember business guru Tom Peters' (1994: 200) recommendation, 'Hire curious people'.

In a highly dynamic tourism situation, great knowledge and great expertise are required if the industry and its partners are to answer the demands
for special adjustments. It is not easy to be 'light' in approach, especially for
a weighty and complicated group that must make its plans far ahead.
Indeed, the mergers and integrations that are occurring within the tourism
industry might seem to militate directly against the ability to adapt and
respond quickly. However, this concern is largely negated by the fact that
these conglomerates are broken into practically sized units, designed to
cater to different niche markets.

Technology plays a liberating role in making customisation possible. It
facilitates bookings and choices. Notifications of changes can be quickly
communicated to travellers and to industry professionals. The technolog-
ically-enhanced immediacy of the provider–customer dialogue leaves
providers with no excuse for ignorance of the customer's moment-
to-moment needs. The producer is then presented with the choice whether
it is viable to respond to preferences as they happen or to remain more
detached. Of course, cases will arise where, in order to keep a customer's
long-term commitment and loyalty, a supplier will decide to provide a
product to an individual, even though it may not be worth so doing on a
strictly business rationale.

Fluidity and flexibility, with structure enough to operate

The essential overall challenge for the producer lies in structuring events
to avoid chaos. Systems must be put in place to bring proper standards of
product delivery, while also developing strategies and methodologies
aimed at greater flexibility. Industry personnel and representatives must
be chosen and/or trained to fit the task; they must possess the right kinds of
knowledge. The knowledge basis of industry equipment must conform to
the task. When Laws (1995: 152) discusses impact assessment in relation to
tourism developments, he regrets the lack of post-evaluation or monitoring
of projected impacts, and says 'In the absence of that knowledge, planning
remains an imprecise instrument in the goal of obtaining particular benefits
for tourism, and avoiding its more harmful effects'.

Overtly commercial producers, engaged in strong competition, may feel
a certain wariness towards one effective means of gaining knowledge: the
networking process. When suppliers lack a certain body of knowledge
in-house, they may choose to use outside consultants. Public sector and
small enterprises, especially non-profit ones, may be especially well suited
to using networking methods. This is both because they have less reason to
fear sharing with competitors whatever information they have gained, and
also because they may be constrained to find the most cost-efficient way of
collecting new data. The reality is that tourism's complexity and range of
necessary participants (from government planning, economic and tourism

agencies, through the transport, accommodation, tour operation and travel agent/bookings sectors, to community organisations) means that some types of networks and partnerships are almost always present and desirable.

At the practical level, fluctuations in customer demand and changing needs often require responsiveness between providers. For example, when a hotel or bed-and-breakfast is fully booked or unable to meet a particular customer demand, in the interests of long-term good will, it is the practice to recommend another nearby establishment, whether or not a formal horizontal partnership linkage is present. The general template of co-operation is already in place. It should be applied more widely, in accordance with Dynamic Tourism's tenets. Dynamic Tourism naturally calls for the creation of new and untried liaisons, as well as for the strengthening of existing ones, along with more creative and productive use of them all.

Active duties for the tourist, towards product conservation

Information is pivotal to the tourism industry in order to deliver a versatile, viable product, as well as to enable a market to understand the use and care of the item that the industry has provided in response to the market's perceived wishes. If the product is a sensitive piece of landscape, a unique or perishable item or habitat, a fragile culture or host community, a vulnerable monument or antiquity, its chances of surviving undamaged will be greater if its visitor audience is sympathetic and is informed about potential dangers.

One dimension of Dynamic Tourism concerns informed visitor choices, and the acceptance of changes and adaptations in travel for reasons outside personal preference and pleasure. Parks Canada (1994: 17) cites 'Education and Presentation' among its guiding principles. It supports the sharing of information, including published and unpublished results of research and says that, 'The provision of accurate, comprehensive and timely information is important to fostering awareness, appropriate use and understanding'. Informed visitors must take responsibility for their actions in a mature way, rather than remaining ignorant and thus open to acting inappropriately, either inadvertently or through manipulation by people with ulterior motives.

Information Itself

Obtaining and extracting information

To make optimal, reasoned choices about destinations and their circumstances, travellers must have access to a background of specific knowledge, as well as important generalised information about *how to find out*. To address only the on-site dimension is to consider many aspects too late, whether from visitor or industry motives. From the visitor perspective, for example, a site that suffers from overcrowding may have lost its 'spirit of place' (see below), or the aura required for a successful visitor experience. When tourists have prior knowledge of the time that visitor numbers are at their highest, they can avoid disappointment by avoiding those peak hours. On the industry side, the careful study of till receipts at paying attractions and associated shops and eating-places is enough to reveal a mass of information about tastes, timings and usage. This will assist in site care and management and structure operations on the most efficient and cost-effective basis. Such information will also reveal early indications of rapidly impending shifts of patterns and taste, in order to react to them without allowing any gap of understanding to develop.

Reality and variety

Tourists often suffer from a lack of information. We may see golden pictures in package tour brochures, for example, but the necessary detailed information often fails to accompany those gloriously appealing images. The situation is changing. It is an interesting exercise to try to analyse a brochure for its intended target market and that market's educational level. The noticeable general trend is for more information in material that is targeted at the well educated. Of course, this is not the whole story, since educational level is not the only target factor, and an audience can be further analysed according to the degree of visual or verbal literacy. Also, we want and expect travel to stimulate us, to be an escape, and it would be an unresponsive travel industry that did not 'talk' to us in those terms. Frequently, these fundamental features are most potently communicated in pictorial form.

Sources and layers

The tourists of the past were often not very well versed in tourism; we accepted the 'baby talk' directed at us by an industry that perceived us as travel infants. Nowadays, the industry must provide the more sophisticated tourism product demanded by seasoned travellers, and it must have data to match its product. As in other areas of society that are now more dynamic and changing, the information in question must be continuously updated; it cannot remain static.

There are ways for the industry to achieve this transformation. Technology has already been described as a great enabler. The Internet is an established source of continually revised information and buying capacity; its myriad sites vary from such giants as Microsoft's *Expedia* to the sites of an individual operator, place or venue. Not only does the Internet provide items aimed directly at potential tourists, it is also useful in obtaining information on related topics, such as current weather, political events and crises. Why, we can even use it to purchase guidebooks and clothing for our trips. The prediction has been made that, 'By 2003, 25 % of holiday bookings will be made on line' (Paton Walsh, 1999: 12). The Internet offers *layers* of information and adds to this useful dimension the ability to explore and make connections among information categories.

The recognition by providers that our information requirements differ by quantity, complexity and depth, is already well established. We can see this, for example, in the displays of the attuned among museums. Along with this recognition, we need the understanding of how we are most comfortable in absorbing information, and so whether we prefer to receive it as written or spoken word, or visual image. Providers must also know where prospective travellers are most likely to seek data, whether over the Internet, on the digital television screen, in an advertisement in a broad-sheet newspaper, or through a travel agent or a tour operator's brochure. A further dimension for providers is to know us sufficiently well to understand what resonant words strike chords of response within us, because of our identity as people and tourism consumers.

The stratified approach to information is evident at many attractions and venues. It recognises a range of duration in time when the visitor is open to take in data. Its questions are these:

- How long is our visit?
- How much of that total time is chosen for absorbing information, and how much for visiting the site shop, café, or lavatory?
- What is the extent of our capacity of data absorption?
- What is our educational level?
- How great is our ability in the chosen media used at a site?
- How great is our interest in the site's subject?

Most likely, the style of media used in reaction to a visitor's perceived comfortable communication method will deliver a certain style of response in the way that same visitor uses an attraction. A mass of interactive screens and myriad features aimed at short attention spans will be likely to encourage visitor action that is hasty and 'busy'. When listening devices such as earphones and wands are provided, they leave the visitor in an individual bubble of concentration.

At Stonehenge, for example, listening wands are used with a pro-gramme that is available in nine languages. The visitor is offered portions of information keyed to accommodate the visitor's available time and desired amount and breadth and depth of information. Visitors, as they circulate the stones in the company of the first and main piece of audio information, are thus led to make a measured, orderly and reverential visit to the stones, because they are focusing on words close by. Doubtless it suited the management, for control reasons, to elicit this response. Nevertheless, the method also seems chosen to tap into the clear spiritual need and desire for individual experience that Stonehenge inspires in many of its visitors. There are various other sources of information available for purchase to accompany a visitor to Stonehenge. These items are in the site shop; they serve either as adjuncts to the data offered on the wands, or as replacements for them, while also fulfilling the role of souvenirs of the visit and sources of post-visit information. The Stonehenge audio wand visit is an example of several dimensions explicit or implicit in Dynamic Tourism: the solitary walk, the feature of spirituality, and the individual and 'deep' tourist experience, as opposed to a corporate and superficial one.

Styles and connotations

Spirituality seems to be gaining recognition as a feature that we seek among our tourism experience options. Places have long been recognised as endowed with the capacity of conveying their own particular *genius loci*, their spirit of place. Such a spirit can then engender in us a response of inspiration, contemplation, calm or stimulation. The advantage for tourism promoters is that almost any destination, be it town or country, a religious or a secular site, can be presented as demonstrating a spirit and/or capable of engendering spiritual response in its visitors.

More remote and unfrequented rural areas may find the spiritual aspect an especial blessing. Quiet, tranquillity, solitude and lack of modern features may be promoted as indications of dimensions of spirituality or religious inspiration that are buried or lost in the urban centres of the contemporary world, except for the remaining places of worship to be found in those cities. In France, the Association for the Promotion of Museums in the Auvergne uses its leaflet 'Every emotion Musées d'Auvergne' to boldly proclaim, 'The Auvergne invented spiritual values' as it promotes its regional museums. The pamphlet continues in a poetic and romantic vein:

> The Black Virgin and pilgrimages under the sullen heat of an August sky. Fresh springs where an anonymous hand has left an offering. The healing virtues of flowers and herbs, and ancient country folk patiently recounting their tales. The Auvergne invented the silence of the shepherds bringing their flocks back to shelter, and the gentle chime of cowbells echoing through the stillness of the evening. (Association for the Promotion of Museums in the Auvergne, leaflet)

This piece is redolent with characteristics of Dynamic Tourism. It illustrates the power that comes from choosing the *mot juste* for a specific audience and task. Here the style of language and words used are among the tools used to impart information. These stylistic choices are especially important in conveying mood and spirituality.

This extract also evokes another important method of communicating information, the *spoken word*. Guided group tours and visits, conducted by a knowledgeable person, are traditional, even routine. In some instances, these group experiences can be disappointing for the listener since, when a whole group is addressed, no individual is catered to in a particular way. This form of visit can represent a lazy option for the provider, leading to a standardised, superficial sort of patter being delivered to the tourists. Oral history or retelling, however, is usually a dynamic experience; the presenter of information is either directly involved as a participant in the experiences of the story, or else is intimately connected to the story in some other way.

By means of oral histories and other living narratives, the storyteller and audience are allowed to communicate outside the written text. A group that might otherwise be silent, its message left unknown, can have its say. Furthermore, recordings of oral performances can spread these stories, retell them and deliver them to a wider audience, scattered much further afield, than in any one-on-one session.

Consider the example of a leaflet from Qantas and the New Zealand Tourism Board that illustrates the potential charm of such oral experiences. It is characterised by a lightness of touch and evocation of spiritual appeal. The 'sell' is soft, and in the absence of 'hard' marketing it manages to imply that the market for New Zealand, potential customers for Qantas, is visually literate, averse to receiving a mass of heavy promotional data and unlikely to need it, since these individuals are presumed to already be quite knowledgeable about New Zealand. This market is reachable by means of connecting with its deeper inner dimensions. Offering 'A few words about New Zealand', the advertisement presents each word alone on a page with a single image, 'heavenly', 'pilgrimage', 'nobility', 'introspection', 'euphoria'. This sophisticated presentation of an aura is destined for a sophisticated, adventurous, individualistic audience, somewhat a cut above the crowd.

Changed ways and attitudes of approach to data and their absorption

Understanding a market and communicating with it at the level of knowledge illustrated above is an exercise for the skilful supplier. If we are becoming Neo-Nomadic, as this book projects, it could be expected that we would develop ways of absorbing data and 'seeing' that differ essentially from a simple progression and variation on existing styles. Marshall McLuhan (1962: 64), while discussing the various ways mankind has used its senses through time, makes the logical statement that, 'A nomadic society cannot experience enclosed space'. This remark is the tip of an iceberg, whose whole circumference we may need to explore if we are making the radical change to a nomadic life. Accordingly, there is a lesson here for the tourism industry: it needs to know us as markedly altered. Consequently, the industry must communicate with us in a substantially different way.

As members of society, we must now accept the obligation of keeping ourselves informed on a continuous basis, in the realm of tourism as in everything else. As providers and consumers, we need to be knowledge-able and in touch. Patullo (1996: 71–2) reports how, in order to arouse awareness in the resident population of how to address tourists, the programme 'Hello Tourist' was used in 1977 at primary school level in St Croix in the Caribbean, and then 1992 in St Lucia (where it was accompanied by an adult programme). In Great Britain the National Trust now espouses the concept of life-long learning through its Minerva programme, launched in 1995, with corporate sponsorship from such businesses as Grand Metropolitan.

One example of the Trust's agenda for schoolteachers is a publication entitled *Learning from Country Houses* (National Trust, 1995), which the Trust has provided with help from Grand Metropolitan. It is commonplace and expected that organisations that maintain and present public attractions, such as these country houses, have an educational aim, an intent to deliver specially prepared informational resources to teachers and school-children, and that facilities directed to that end will be in place for use during visits. The new approach, as has been noted, is to accept that we are all, regardless of age, in a position of learning; we all need to take in information throughout our lives, not merely during our formal education.

Dynamic Tourism specifies our need to be informed, to be aware, and to make informed choices. As tourists on day trips and holidays, we are in the mood to enjoy ourselves, we expect to have fun. As travellers on business and educational visits, it is still our human nature to prefer that our visits be pleasurable. From the provider's perspective, independent of informational reasons, it is necessary to convince visitors that they are having a happy experience, so that they are in the frame of mind to spend money during their present visit, and so that they are also encouraged to return to that particular venue on another occasion. In a theme park, for example, where the essence of the experience is enjoyment, there is a particular style of communication that is now being transferred to other situations, where education is the first objective.

For example, the matter-of-fact, dry style of exposition that characterised many museums in the past, has been recognised to be far from appropriate for establishing contact with many audiences. The style of communication selected by a provider should recognise the various dimensions that we have mentioned previously. It should be chosen against a general background of a perception of where information will be more likely to be absorbed and remembered by the recipients if the circumstances rendered and the method used are interpreted by them as pleasurable. Another aspect of '*edutainment*' that is cited as a reason why this form of presentation has 'taken off' in the UK is that we now want 'value for time' and 'if time is at a premium, then people want to do or learn something at the same time as making whoopee' (Jones & Stuart, 1998: 2).

The provider–consumer dialogue

How are providers choosing to communicate with visitors in ways that may be related to Dynamic Tourism? The essential necessities are to *connect* and then:

- To set out options for choice and to make them accessible and understandable.
- To provide adequate further information for responsible and enriching activity.
- To ensure a capacity for change.

We have seen that technology is a great enabler in the search for knowledge. Technology helps in the provider's need for flexibility and efficiency, in planning, managing and monitoring. Handling reservations, staffing, ordering and purchasing are all made easier this way. In the attempt to suit the aims of both provider and visitor, *what* is said is important, and so is *how* it is said. As we have already shown, the medium of choice is the one best suited to the target audience.

On location

On the ground at a tourist site, the first necessity for the provider is to orientate the visitor and, as during the pre-visit planning stage, to offer choices. Tourism information centres and various types of site services cater to visitors in this way, often fulfilling this function at national or regional entry points. For example, consider the traveller resource complexes found on French Autoroutes. These centres provide the practicalities of food outlets and picnic areas, toilets, cash points and petrol stations, but they also offer a shop with local crafts and foodstuffs for sale, and a tourism information centre, often with some form of video or slide presentation.

The Volcans d'Auvergne Services area is an illustration, situated at a gateway to the volcanic hills of the Auvergne, and at a good scenic lookout point as well. This centre also has a hotel within its perimeter. A large relief map of the area is located in the visitor centre and offers the tourist options, as would be recommended by Dynamic Tourism. The tourist chooses among an array of possible themes and views a simple video presentation, linked to lights on the map that illuminate specific locations at the appropriate moment. It should be mentioned that slides also are a presentation method that can be used in similar situations. The advantage of a slide montage is that it can be altered relatively easily, with individual slides exchanged or substituted, repositioned or re-juxtaposed, or replaced with new transparencies. In contrast, a film or video does not allow such flexibility and can only be altered with difficulty, so that the only reasonable option for an update is complete replacement, at what is likely to be considerable cost.

Reasons for Communication

In Chapter 2, we mentioned the firm educational role of tourism information, in the case of the display on human-caused pollution in the Shenandoah National Park, in the eastern United States. An additional facet to education might be its role in justifying and explaining the use of public money, while gaining ongoing support for projects, such as an archaeological dig. In the United Kingdom, the usual practice is to invite and in some way facilitate public viewing, with the means determined by the site's fragility. In 1999, for example, the Museum of London presented the activities of its dig at Spitalfields. The display included what was, because of its high standard of presentation, effectively an outreach Museum display near the site. Print media resources were also available, with the brochure *Archaeology Matters* (Spitalfields issue, Summer 1999). This brochure responded to the needs of the location in contemporary terms by its bilingual text in English and Bangladeshi, acknowledging the presence of a considerable Bangladeshi community in the area. The underlying assumption in such instances as the Spitalfields dig is that archaeologists and the public are on the same level, rather than engaged in the relationship of 'authority' and 'ordinary mortal'.

As a point of contrast, let it be noted that, in 1999, on the coast of southern France, a divisive, haughty and unfriendly face was turned to the public at the important maritime and religious site of Maguelone. There the public was excluded from the archaeological site and not told *in situ* about the findings, although considerable public money was being spent on the endeavour. Earlier in the decade, at Santa Barbara in California, the happier example of the Casa de la Guerra kept the public well informed about the historical research and archaeological exploration that was occurring there. The public was kept in touch with the plans for interpretation by means of the house and a museum, 'for the community and its visitors'.

In the United Kingdom, the policy of regarding the public and tourist as equals and helpers with site personnel has been taken yet further by the Peak Environment Fund, a charity whose projects are meant:

to protect the environment, the traditional land uses and communities
... from the adverse effects of visitor pressures

to improve public access ... where this does not have an adverse
impact ...

to improve public understanding ... of the area and through this to
encourage their support for its conservation and management to
reduce the adverse impact of cars ...

to encourage the use of public transport, cycling and walking ...

to encourage recreation and tourism in the fringe area around the
[Peak] National Park ... (*Peak Environment Fund Background Information*,
undated, but issued in 1999)

Here, it is made clear to Peak District's visitors – of whom there are over
twenty-two million each year in the Peak National Park itself – that they are
the cause of much erosion. The implication is clear that visitors, since they
have a negative impact on this fragile environment, have a parallel obligation, as people of understanding and responsibility, to contribute to
damage repair once they have been made aware of the problem.

Special Messages

The warning messages that the Shenandoah National Park and the Peak
Environment Fund present to their visitors are serious ones, important
enough to communicate to society in general. It frequently happens that
tourism, as a global endeavour that is complex and 'new' in its major form,
will serve to reveal something of more general relevance and importance.
In the United Kingdom, the organisation Tourism Concern acts as one
keeper of the conscience of tourism, while communicating its message. In
1999, the organisation launched a new information campaign. Crace (1999:
10–11) reported in *The Guardian* that tourism's widespread ramifications
were cultural (as traditions are lost or damaged), social (as family, customs
and lifestyles change), economic (with benefits not necessarily filtering
down to the local community), environmental (in disruption to landscapes,
wildlife patterns, water resources and other natural factors) and political
(as tourism is used towards political ends, such as exploitation in
Burma/Myanmar).

Certain areas of frequent difficulty in tourism gave rise to debates held
by Tourism Concern and two newspapers in the summer of 1999, in
conjunction with the University of North London, on whose campus
Tourism Concern is based. In parallel, Voluntary Service Overseas (VSO)
conducted a Worldwise Tourism Campaign, 'to demonstrate public
support for fairer tourism and show tourists how to get more from their
holiday' (VSO, in association with *The Guardian*, undated supplement: 14).
Another organisation, the Centre for Environmentally Responsible

Tourism, operates a 'kite-mark scheme for tour operators who have committed to environmental policies' (ibid.).

Such organisations and activities as these demonstrate how importance it is for society that the shortcomings of tourism are widely communicated. In the past, the public message on tourism has tended to be only the over-effusive or simplistic promotional claims of the industry. Now the industry itself is showing signs of accepting new responsibilities. For example, the World Travel and Tourism Council (WTTC) has inaugurated its own environmental scheme, Green Globe. There have been more recent initiatives on both sides of the debate; they may demonstrate society's underlying belief that issues must be examined, that the general world citizenry now expects to be equipped with information, not kept in the dark. Having more information enables us to make our own choices, and the current climate clearly encourages us to gather it.

As a specific example of the necessity of information, as well as the principle that information must not be seen as a camouflage for difficulties, let us consider the 1995 *Integrated Tourism Master Plan* delivered for Uganda by the United Nations Development Programme (UNDP) and the WTO. The *Plan*, as its title suggests, is inclusive. It identifies promotion as a necessity, to acquaint both the travel trade and travellers with Uganda as a tourism player, and to disseminate contemporary information. Interestingly, one proposition in the endeavour aimed, 'to prepare tourists not to expect highly-developed facilities and tourism infrastructure' (WTO, 1995).

Science is one subject that many in the general public do not find enticing, although we recognise its importance intellectually, and acknowledge that we need to learn more about it. Science is now taking on a greater role in the tourism arena. Traditionally, science has found its place in old-style museums, with only certain enlightened exceptions in terms of visitor appeal, such as the Science Museum in London (see Chapter 2). However, science has now emerged as a visitor subject at less orthodox venues, and in entertainingly interesting ways. Perhaps it is exactly because science presenters know that our attention is often not easily engaged with the subject that their attractions are increasingly among the most innovative, communicative and appealing.

Two examples of innovative new science centres in the United Kingdom are the Earth Centre near Doncaster and the International Centre for Life at Newcastle upon Tyne. That both Centres have received Millennium Commission lottery grants is perhaps a sign that today's public attaches a new importance to science. The Earth Centre, with displays both indoors and outdoors, uses theme-park methods to impart authoritative information about the environment and about eco-friendly, alternative lifestyles. The International Centre for Life opened in the millennial year. It is a multi-feature centre, based on a living research laboratory of information and discovery, The Institute of Human Genetics of the University of Newcastle upon Tyne. Along with the Institute itself, guests may visit the experiential attraction, Life Interactive World (expected to bring in up to 300,000 tourists a year). There is also an Education Resource Centre, and offices for companies engaged in biotechnology. A special dimension for the Centre is its role as a site for debate and inquiry into ethical and policy issues concerning genetics. Thus we see a visitor attraction that is part of an environment at the forefront of scientific discovery and accumulation of knowledge. Because public space is an element in the complex, general public awareness of the Centre and its contents should develop.

In southern France, at the Science Route in the Herault Department, we see the manifestation of a simple idea about bringing science to the people and addressing fundamental matters of space, land and water. The Science Route incorporates a number of institutions: an observatory; a visitor centre in the Natural Regional Park of the Haut Languedoc that addresses such subjects as geology, ecology and climate; Florilab, a horticultural centre; a lagoon ecosite; Agropolis, an internationally renowned university agricultural museum; and Epidaure, a major centre for the gathering of information on, and prevention of, cancer. Many of these elements, as you may notice, are by their very nature, dynamic.

The varied components of the Science Route are weighty. They examine serious subjects and assume their audience to possess an answering serious and sophisticated interest in science. Their important dimension, however, is that they do not confuse seriousness with a lack of attraction. They are approachable and they use sound methods of communication, making the experiences they offer to the public pleasurable and appealing. An additional, but essential, dimension of the Route's formulation, of course, is that the target public should know of its existence. We must emphasise that it is vital for a tourist attraction that information about it be put before tourists at places where they congregate, such as airports, railway stations and tourism information centres.

Key Overall Aspects

When we take an overview of tourism's effort to gather knowledge, what key items emerge as significant? One clear item is revealed in the science examples just cited: providers must not patronise their visitors. The contemporary, experienced tourists who are the target market for Dynamic Tourism presuppose that they are to be treated as equals. Equality of access to information is also of paramount importance. As Charles Handy (1994: 201) advises, when he speaks about intelligence (in the sense of information), 'If we don't make this new property more widely available, if we don't invest in the intelligence of all our citizens, we shall have a divided society'.

Dynamic Tourism delivers the means for tourists to make their *own* choices and decisions. To provide information and make it attractive enough in itself for the tourist to pay attention to it and keep it in mind, requires the provider to concentrate on the tourist. Such concentration includes the minute understanding of the form in which consumers prefer to receive data of different types, on varying occasions. The message must be focused in its appeal and presentation, with a focus that decides what media or combinations of media should be used, whether in words, visual images, sounds or touch, or perhaps even by an invitation to the visitor to try to 'feel' through role playing. The style of the message, its degree of complexity and depth and length of delivery, must all be decided for optimal impact.

As we have noted, one of the necessities for Dynamic Tourism is for ways of delivering change to be 'built-in'. We must seek out the most adaptable possible means of communication. Creative ideas are clearly a core requisite for publicity to be sympathetic, suitable and original enough to attract and keep the attention of the tourist audience. The overall notion need not be complex. The most flexible ways of delivering information can be the least expensive, in terms of technology and materials, although some may be very expensive in terms of human resources. The essence of the Dynamic Tourism message is simplicity and adaptability. For example, it is very simple and adaptable to deliver information by talking. The linked necessity is that time, trouble and skill must be spent in adapting the customised message to suit each audience.

Are the data delivered to travellers as part of the tourism process 'education', 'learning' or 'knowledge'? 'Education' and 'learning' may carry elitist connotations of worthiness or superiority that may work against the effort to deliver information. Such semantics may seem minor, a distraction. However, all possible connotations of publicity materials must be analysed in order to avoid any impediment to the effective transmission of information. As we move toward the precepts of Dynamic Tourism, the important mission is to put the individuals engaged in tourism in a situation of choosing for themselves from a position of knowledge, so that they may show mature responsibility about information and its use.

Chapter 4

Going

Context; and Potential Causes to Go and Not to Go

Going on holiday is normal for many of us now, and is expected to become so for many more people in years to come. We like to travel, and the travel industry likes us to like it. We are encouraged to take our travelling as a matter of course. However, we also recognise reasons not to travel to certain places, in certain volumes, at certain times, and not the least, because more available information tells us so. This realisation comes to us as members of society, as well as travellers. Matters inhibiting us might be distaste at political, social or cultural circumstances surrounding a particular travel option. We may be deterred in relation to a venue's 'carrying capacity' in spiritual or physical terms. There may be economic reasons that affect the cost, value and affordability of a destination, or there may be problems related to levels and presence of such physical factors as booking opportunity, resources of accommodation, access to transport.

The idea of travel is thus fraught with pros and cons: will we or won't we make that decision to pack our bags and go? Such aspects are the accepted background to Dynamic Tourism. This chapter considers the issues involved in the process of choice. When and how, in the general context of *choice* is the circumstance constructed that is regarded by the tourist, and created by the provider, as suitable for travel? This discussion will seek to outline the necessary characteristics and conditions, whether pre-existing or created, for the adequately informed individual, placed in the position where a decision must be made, to feel able to travel and to expect to feel comfortable in doing so.

Demand management on the supply side influences the outcome of deciding to travel or not and, if so, when. Demand management reflects seasonal variations in tourism: different price rates are used by suppliers in relation to these variations in order to influence visitor patterns. Of course, this is a customary way for the tourism industry to manage and spread visitor use of venues and facilities. However, if this manipulation continues to be only a deployment and arrangement of visitor usage of the *same* products, no help is given towards opening more destinations for the tourist to seek out and use. Still and all, as new marketing directed to today's dynamic tastes emerges, more low- and shoulder-season travel to existing resources ought to be possible, thus increasing the volume of travel.

How Dynamic Tourism Relates

The dynamism of Dynamic Tourism is the key to providing for ongoing travel. Travel and tourism 'going' are the objective, and Dynamic Tourism's features of extending and encouraging fluidity, of routinely embracing and accommodating transitory situations, should become part of the attitude and practice of the tourism industry.

With short attention spans encouraged by modern media such as television and the Internet, and with a general appetite for novelty and fresh excitement, today's consumers will have no difficulty in making the change; in effect, they are hungry to do just that. To us, tourism is relaxation and/or stimulation. Even business tourists have the same instinct, to have an experience characterised by freshness, interest and enjoyment, as much as work commitments permit. We are looking for a pleasure zone. We are willing to encounter it in many and varied places.

The tourist industry itself faces a considerable challenge alone, the practical difficulty of organising and structuring circumstances that will produce operational efficiency and financial viability, while visibly delivering more flexibility. It may be less easy to make money in the new situation, or at least to do so directly. It may be necessary to find new, more thoughtful, less formulaic ways of attracting income. Because of such considerations, we must not underestimate the difficulty of the fundamental changes required of the industry and its partners. This book's argument, however, is that in the near future these necessary alterations are likely to be unavoidable on commercial, economic, societal and pragmatic grounds.

Monitoring and analysis in preparation for shifting and changing

In order to get away on a trip, tourists need an extended portfolio of options, and flexibility in the use of these options. On the supply side, use of resources by visitors should make it possible to be variable, to reflect immediate circumstances. Is erosion of a natural site reaching an unacceptable degree? Is it then advisable to reduce the number of visitors, or to prohibit visits entirely for a while? An individual site's decision for action needs to be made in the context of the overall situation of tourist venues. The venue portfolio should include the provision of a wide number of temporary attractions, a chance to exercise Dynamic Tourism's criteria of immediacy, flexibility and relative spontaneity. Such temporary attractions can be provided reasonably quickly, and thus help to relieve the overload of visitors on other sites.

Here it is immediately visible that monitoring tourism (for example, assessing its impact or the effects of new introductions) is very important in providing the necessary information on which to act, and from which to respond. As we have said, the monitoring process is currently a weak spot in the tourism industry. It must be developed, however, as an overall aid to implementing new recommendations. Monitoring depends, obviously, on new cyber-technology. The huge capacity of computers for absorbing data from a range of sources, integrating them, ordering and re-ordering them and analysing them by different criteria and categories is what will make the recommended new overall tourism practice possible for the tourism industry. The result will be to make it easier for an increased and altered tourism from the tourism industry offer to the public.

Seeing and finding more sights

One important thing about our general ability to go on trips is that we should perceive more kinds of sites as tourism venues. Often the only required change is that we should alter our perceptions of specific places, since certain sites that we now might recommend as destinations for a visit have always been objects of visitation, but have not hitherto been seen as specific goals for *tourism*. Often they are items that we include as part of everyday life, but that we nonetheless treat and regard as pleasurable activity. Maybe they are among the many sites that have suffered from visitor and tourist industry inattention, because they just weren't perceived by either as attractive. Maybe they just haven't received enough promotion from the industry for people to notice them, or the publicity they have received is unsuitable or inadequate.

It is essential that the tourist purview should be allowed and encouraged to include new places as within the realm of tourism. In 1995, Urry recommended:

> We should get away from the tendency to construct the tourist gaze at a few selected sacred sites, and be much more catholic in the objects at which we may gaze. This has begun to occur in recent years, particularly with the development of industrial and heritage tourism. However, in part the signposts are designed to help people congregate and are in a sense an important element in the collective tourist gaze. Visitors come to learn that they can congregate in certain places and that is where the collective gaze will take place. (Urry, 1995: 139)

Tourists' choice

Urry's remarks are already interesting, but when we take them a bit further they become even more so. The third sentence, in particular, needs 'turning on its head'. We need far wider and radically different products, because a more radical extension is needed than that which Urry observed had begun to occur in 1995 when he was writing. The fundamental change noticed and heralded by the theory of Dynamic Tourism is the growing accomplishment and sophistication of today's tourists. These aware people reject being corralled and steered by the tourism industry. A larger segment of the market, not just the minority 'educated sector', resists the general 'cattle call' marked by the use of standard, routine products. With more travel experience than our parents, and with better information about possible destinations, we feel sufficiently confident and motivated to make informed choices. We assemble products and journeys in our own right, for ourselves. Our choices and product assemblies reflect our personal desires, rather than being hewn from the narrow, existing mass templates put before us from the travel industry up until now.

Access and ownership

So then, what mindset are we cultivating, and what sort of venues do we, as tourists, need to envisage as attractions in the future? One important priority is to be sure that the new expanded number and range of destinations is *accessible*. As the industry promotes more numerous and wider-ranging goals, visitors must be assured that they are welcome; they must be admitted and encouraged to enter the destination and areas around it, and the locations themselves must be accessible to transport. The people who own or control goal territories must be enthusiastic about extending access to potential visitors, or the necessary enthusiasm must be persuaded to develop. It may even be necessary for legislation to ensure reasonable access to new attractions.

The question of persuasion is related to the knowledge issues discussed in Chapter 3. A campaign of information to area landowners may be the necessary instrument that will cultivate in them a favourable frame of mind and encourage them to see that, when they allow increased public access, they show both good citizenship and good public relations. They may also realise that allowing access for tourism, which helps the economy of their surrounding community, might also directly represent good business for them in their own right.

The issue raised, that of distinguishing private space from public space, is a hugely important dimension of the development of new tourist venues. When a country's law allows exclusive property rights, major distinctions can develop. One must distinguish between parcels of land in individual ownership, depending on whether the right of outsider exclusion is exercised by particular owners or not; and one must distinguish between those private parcels and land that is owned or administered by public sector bodies on behalf of communities and society in general.

These differences will greatly affect decisions on the fate of a given site. Whatever the type of ownership and approach characterised in connection with a certain place, those other uses made of the territory further affect tourism provision and attitudes to it. For example, if farmland is actively cultivated, or if game shooting is carried out on it, the status of the land affects its accessibility for tourism. In the case of cities, some of the space of city streets and squares is needed by shopkeepers to receive deliveries of goods; this space may not be fully accessible to the demands of tourism.

Compensations and alternatives

Three factors influence a destination's viability for tourism: the character of the venue itself, the transportation network leading to it, and the range of accommodation available for visitors. In the case of a day trip, only the first two factors are relevant, so if a site is characterised by a shortage (or even, absence) of accommodation, the traveller has the option of returning home at day's end. If a preferred type of transportation is unavailable, whether it is too full to accommodate more passengers or is non-existent, the choice remains to use another form of transportation, if practicable. Cycling and walking are usually possible, except for people who suffer from physical disabilities.

If an established venue is full, out of commission for repair or 'rest', or has become ethically unavailable for a raft of possible reasons including political unrest or open hostilities, then we must look for alternatives. When a popular site must be closed, the tourism supplier, in order to offer other options with the same or comparable benefits, must analyse and understand what motives drew travellers to the original site. When suppliers are trying to produce alternatives to over-visited or fragile sites, an evaluation of existing tourism possibilities that do not seem to suffer from the same stresses of overload or breakdown can produce leads on the best alternative tourist experiences to resolve the problem.

Re-introductions and New Additions

Former attractions, that have gone through an eclipse but now are refreshed and extended, can offer one set of possibilities for continued tourist travel. The same is true for altogether new attractions and newly recognised existing sites with potential for tourist exploitation. Chapters 5 to 8 will focus on these possibilities for expansion of tourism, so they will not be treated in depth here, although there is some overlap with the subject of the present chapter. This chapter's role is to look at, and for, non-traditional venues and tourism activity.

Perceiving and bringing in new products

The groups of countries and regions that are not overtly active destinations in contemporary tourism are a large potential source of fresh tourism products. While some of these areas may deliberately avoid tourism, others may very much wish to be visited, for economic and other reasons. The cause of their omission from the 'tourism map' thus far is probably either a function of insufficient infrastructure development, lack of basic amenities, or inadequate publicity. Another reason is that they may simply not have fitted the old profile of appealing destinations.

As Dynamic Tourism suggests a change of taste and attitude in its premise, as well as the assumption of more individual responsibility for choice on the part of tourists, new opportunities arise for these destinations. However, these places that are not in the swim of tourism may be out of touch. Key people there may simply assume that today's tourists would reject their product, or demand a standardised format and massive material development in their facilities. It is necessary to convey to them the realisation that things have changed; the tourism situation has altered and is developing along more flexible lines. Then, of course, the need for them to reciprocate and convey an awareness of their products to us becomes paramount, with an importance that increases with time, as we have seen in Chapter 3.

Large portions of Eastern Europe could be suggested as examples of under-visited regions that are either insufficiently prosperous or insufficiently developed to present a tourism product that conforms to the traditional format. People there may be unaware that their existing assets and features should already appeal to a dynamic tourist. Such places may present rudimentary facilities, but their very defects, in terms of old-style tourism, may be used to advantage and serve to attract new travellers. After all, the developed world is lavished with material goods, and the essence of tourism calls for a contrast to the routines of daily life. Visitors from affluent areas, who seek the features portrayed as part of Dynamic Tourism, may see considerable charm in their polar images, finding attractive the globe's 'backward', undeveloped or under-populated areas. To appeal to consumers for some of the same elements as these underdeveloped areas are those products that manifest simplicity as their key and defining feature. And often alongside or as part of these items are those that deliver a spiritual dimension.

In trying to gather experience, rather than stockpiling material, tangible and fixed possessions, we are rendered as having Neo-Nomadic characteristics. In regard to tourism, we relish those aspects of life that 'first phase' travellers of the modern day would neither have noticed nor prized as tourism products. As a consequence, items that were not 'in the frame' as tourism products can now develop a market alongside more conventional offerings. Consequently, the menu of tourism options is expanding.

As people of today become more sophisticated in their outlook, they may see an attraction in tourism articles that are abstract as well as figurative. Perhaps this attitude parallels the spiritual or religious impulse, which we have already considered, in that each of these dimensions relies upon our accepting the proposition that things need not all be 'spelled out' and literally defined. An appreciation of abstraction would finally bring into wider recognition characteristic features of Modernism, so these too can be offered among a greater overall group of tourism products. When the Bauhaus and its sites at Dessau and Weimar were designated as a World Heritage Site in 1996, signs of the possibility of such a development became apparent. Since these sites are located in an 'untouristy' area of the former East Germany, they offer a useful dimension of expansion for travellers.

At the other end of the time scale, an increased consumer appreciation for abstraction and a greater ability to cope with the absence of clearly defined legible meaning could result in the development as destinations of a huge number of attractions that are traditionally 'poor relations' in the tourism market. These destinations are the simpler monuments of prehistory. Stonehenge stands to date as an honourable exception to a general apathy among tourists towards these sites, so typically austere and devoid of ornament.

Substituting; and Authenticity

Yet another possibility for expanding the number of attractions available to the multitude of tourists wishing to get away is the immediate substitute. Such alternatives should be given more attention. They may often be man-made replacements of an original item, whether it is unique, historic, or a natural feature. A museum or visitor centre may be a substitute for the 'real thing' that a historic feature represents. Its success in serving in this role depends on its capacity to deliver, at least to a significant number of visitors, a vicarious experience that is as good as or better than the original. And tourists may better accept the substitute when they understand the worthy aim of using a substitute so that the original may be better preserved.

One factor that governs success in diverting tourism through substitution lies in how much a provider's chosen tourist market prizes authenticity, in the sense of contact with the original item, rather than with a good reproduction, as in France's Lascaux II Cave, for example. Since the substitute can be fabricated to order, in the highest quality, while with the original the only hope lies in the best preservation and presentation of an existing product, the task of effective diversion should be achievable. Moreover, in the case of substitutes, providers have the increased possibility of adding on resources and moneymaking facilities, thus increasing the substitute's attraction to visitors. Indeed, the method of replication makes it possible to deliver a specific tourist experience to an audience far removed from the original site. Thus, there is an added benefit: the location of a substitute, replica or extra can be chosen so as to bring tourism to an area that will welcome and accommodate it, rather than leaving the whereabouts and circumstances of an attraction to the dictates of chance in the location of the original site.

A substitute that enables us to 'visit' a particular attraction may thus be either a replacement, or an addition. For example, a replica of the Berlin Wall was built for display in a Florida theme park (Rowinski, 1995: 14). This replica presents a far-away addition, but its appeal in relation to the original is affected by the fact that very little at all of the real Berlin Wall remains, and even less of it is still to be found *in situ*. In architecture centres, another form of substitutes, visitors have the possibility of experiencing inaccessible monuments, where the originals are in private ownership and not open to tourists. Architecture centres also gather items from various places and arrange them by theme, as do so many art galleries and museums, thus presenting extra attractions for tourists who are also visiting sites for a view of the 'real thing'.

Artificial by nature

Of course, there are numerous tourist attractions constructed precisely to entice visitors, not as replicas of other hard-to-reach or delicate original entities. The Disney theme parks are large examples. Unlike replicas, which must be planned in the nearest possible relation to the proportions of their originals, these theme parks can be built and altered to accommodate as large an audience as desired. After all, they are new creations and originals in their own right, and need conform to no model other than their own integrity of concept. Of course, this may be said only with the provisions that safety requirements are met at these attractions, and that their impact on the surrounding environment should be benign, including the impact of the transportation that conveys visitors to and from their portals.

A Part of Daily Life

Now let us turn to the large number of features and activities that make up part of our daily life, and from which we obtain pleasure. This group of attractions also features in tourism, although in the past its elements have in themselves only infrequently been seen as reasons to travel, rather than adjuncts to travel directed toward another central object. This rubric includes activities such as strolling, walking and viewing, eating and drinking, discretionary shopping, and experimenting with travel on unfamiliar or rarely used types of transportation. Often these discrete activities are experienced in interconnected ways, where the whole proves to be greater than the sum of its parts. Each individual category contains a huge range of existing and potential options.

Activity, and old styles of gentle recreation re-visited

For example, let us consider the complex of issues related to walking. Among these issues will be the increasing number of products that already exist, or are being developed, to deliver adventures and/or extreme testing experiences to people. For example, tourists can go on treks in the Himalayan foothills, or enrol for a contrasting tour, the more gentle and sybaritic French walking vacations, footing it from one hostelry to another, while baggage is separately transported.

Walking in town, a less physically demanding sort of activity, invites us to consider urban public parks, under-utilised resources often not recognised as tourism products. These city parks and the land within their borders offer a range of inviting features worth walking to see. They provide space to introduce temporary items or further attractions. Since they are located in population centres, they are easily accessible for out-of-town visitors, as well as to large numbers of nearby residents. Moreover, many of these gardens possess the added attraction of age, often dating from the late nineteenth and early twentieth centuries. Chapter 8 is dedicated to the rehabilitation of formerly popular tourist venues, and this aspect of the public parks will be treated in greater detail there. However, tourism professionals are often slow to realise that the elements of contemporary version and style embodied in many of today's parks are enormously appealing.

Consider these examples: the Parc André Citroën in Paris and the Zoo and Botanical Garden of Hong Kong Island. The Parc André Citroën has shown itself inviting both to travellers from out of town and to local visitors. The Hong Kong Zoo and Botanical Garden provides an unusual space, an oasis and site of tranquillity for the Island's residents and visitors, while simultaneously acting as an overt educational resource. Yet another proof of the appeal of modernity, as it was defined in the mid-twentieth century, is the Delegates' Patio by Isamu Noguchi at the UNESCO headquarters in Paris (Altshuler, 1994: 66–69). A part of the attraction of the site lies in the Japanese Garden in association with and alongside the Patio. These UNESCO examples are not the easiest places to access, yet they are still well visited. A Zen character permeates them, and since so many people find them appealing, we may infer there to be a general inclination among many members of the public for this modern style of attraction. Most notably, in the Delegates Patio we may observe the taste for the abstract, which earlier passages of this chapter noted to be present and likely to increase among contemporary travellers.

Sculpture to attract

The Delegate's Patio at UNESCO also draws attention to sculpture, another under-exploited attraction for visitors. Visitors journeying a bit south west of Disneyland to Two Town Centre at South Coast Plaza shopping mall, at Cost Mesa, California, can visit a 1980s Noguchi garden, *California Scenario*, where the sculpture group, *Spirit of the Lima Bean*, is the most notable feature. This sculpture garden has brought fame to its site. It offers a sophisticated addition or alternative to southern California's other attractions. However, even the Disneyland Hotel is graced with a Japanese garden, as was mentioned in Chapter 2.

Appeal to the senses

Our attraction to scents and smells finds its expression in the popularity of gardens in their usual form of land planted with bushes, trees and herbaceous plants. However, the sensory dimensions offer further opportunities for defining other attractions besides traditional gardens. The pleasures we derive from the scents and smells of the natural environment of the countryside are open to be embraced and used more intensively in tourism. Walks and activities could be designed to focus on the senses, joining the olfactory attractions in alliance with the auditory through sounds such as bird songs and the tinkle of sheep bells. In urban areas, the sensory realms can still offer adventures by linking scent to taste through the medium of food and wine consumption.

Drink is beginning to win recognition as an attraction in tourism, with wine tastings, dedicated wine tours and encouragement, through roadside signage, for tourists to visit vineyards. This all augments the whisky trails already in place, and tour tastings of port, sherry and Madeira offered at bodegas and cellars. However, many more venues could be offered. The Vinopolis complex, a major visitor attraction introduced in 1999 to London's Southwark area (to be discussed in Chapters 6 and 8), includes the Odyssey layered-level presentation. Here, visitors take a self-regulated wine-tasting tour while they are told about the wines they experience as part of a CD-audio-informed commentary, and some wine tasting is included as part of the admission price to the attraction. The same site offers a range of restaurants, as well as a shop operated by a wine chain and outlets for the sale of wine associated items, including upmarket 'real' cheeses.

Foodies and imbibers

Food types among tourists are catered for and promoted to at food-processing factories, dairies and certain farms that permit visitors. Visits to such sites are accompanied by the possibility of purchasing the produce, as well as observing its preparation. Even picking fruit in season, an event similar to these visits to the point of production, is now regarded as a *tourism* activity – a rare point of view in the past.

However, once the activity of picking fruit is seen as a personal interaction with an important element of country life, as well as a fascinating phase of food production, and information is provided about the crop and how it is grown, the attraction for tourists becomes plain. Add an on-site café, and the fruit-picking station becomes as much a tourist destination as any other.

Certain restaurants and hotels can become tourist objectives in themselves. One has only to think of London's River Café, Harry Ramsden's various fish-and-chip emporia in the United Kingdom, the London Ritz and the Pera Palas in Istanbul, to name a few examples. These sites of hospitality blaze a trail for many more comparable places to be promoted by their owners and accepted by the public as outright destinations, rather than being treated as 'side-shows' and taken for granted as incidentals. As Chapter 2 suggests, the market of interest in exploring the nature of food and drink, as well as their cultural and geographic variations, is so huge and their fascination so great that they are manifestly under-exploited as a resource for tourism.

Shopping

Several of the foregoing topics imply that an important feature of this kind of tourism is the opportunity to shop. Shopping possibilities are especially necessary for an attraction whose main element is cost-free, whether through necessity or choice. The shop, café or restaurant on offer at the site needs to deliver supporting revenue.

Shops may be destinations in their own right, as is Harrods in London, now almost essentially *only* a shop for tourists. We have, of course, already mentioned the satellite Harrods shops installed at Heathrow and Gatwick airports; they serve to confirm this point. Speciality shops are particular attractions, often luring visitors over long distances to purchase their goods. Many more such outlets could take up a role in contemporary tourism, if their owners so chose. A distinctive shopping mall can act as an expanded destination for shopping tourism; prominent examples are Canada's West Edmonton Mall, the Mall of America in Minnesota, and the United Kingdom's Metro Centre and Bluewater. However, any one of these shopping meccas, if they are not 'refreshed' and refurbished, can be quickly submerged and replaced in the visitor focus of attention by the next, most promoted arrival upon the scene.

Appealing to basic propensities

As we have just seen, tourism providers who cater to us as sensual beings and consumers, by gratifying our natural inclinations, can widen the menu of tourist expectation and experience. Thus we can choose from a greater range of destinations for our travel plans, increase our travel opportunities overall, find alternatives or additions to existing sites that are closed or restricted to visitors as a result of resource overload.

Transport as the direct focus

We are familiar to an extent with models of certain means of transport becoming direct tourist attractions, rather than remaining merely the method by which visitors reach a destination. Old, closed railways are probably the prime example of this phenomenon, but vintage and classic car collections demonstrate a kindred appeal, as do historic and redundant ships such as the Tudor *Mary Rose* in Portsmouth's Naval Docks. Boating vacation possibilities already range through a gamut from canal boat and yacht flotilla holidays through to voyages on cruise ships, with a recent sharp increase to be noted in the last category.

It is clear that tourists already show an inclination to choose holiday options based on a specific mode of transport. Nonetheless, there is still potential for tourism to focus even more sharply on transportation as the essence of a holiday. Like food and drink, transportation is a fundamental necessity for human existence, and it follows that most locations have potentially attractive transportation offers that they can develop, thus increasing the total number of tourism destinations. It should be remembered that part of the attraction of different methods of transportation rests in the spaces they employ and occupy: examples are railway stations and lines, boating canals and docks, airways through the sky and airports on land. Tourists experience unusual views and insights when they are open to these perspectives and insights. The success of the British Airways London Eye, a large slow-revolving spectator wheel (operated by The Tussauds Group) in central London to celebrate the millennium, manifests how a different view provided from a particular transport type can render central appeal.

A different type of example is the National Trust's restored Victorian steam yacht Gondola, which commutes between venues on Lake Coniston in northwestern England. It is an overt tourist attraction in an area already overcrowded with visitors. However, the concept behind the Gondola, which combines the nostalgic attraction of a trip on a historic watercraft with unfamiliar views of scenic locations, is easily transmitted, and could be used to open up other places to tourism. The Sydney Harbour Cruise is another example; although the boat is not historic and so the boat trip lacks the nostalgic charm of the Gondola rides, the striking water views of the city offer sufficient appeal to make a great visitor attraction. The Sydney cruises are specifically focused on tourism development. They are built on the notion of the pre-existing network of ferries that link the communities of the large Sydney Harbour. Among other striking views, the trip provides

Unusual transport, unusual perspective: the British Airways London Eye spectating wheel, Waterloo, London, England

a distinctive panorama of the flamboyantly modern Sydney Opera House – certainly the best possible without taking to the air.

Lightness

Temporary, and less tangible

If tourism is to be flexible enough to respond to changing patterns, contemporary providers must focus on less fixed and monumental attractions than those of yesteryear. These sites should now be brought into focus as tourism features, adding to the attractions already in existence, in order to increase the overall number of available tourism possibilities. In this category of temporary destinations, we may consider the mass of one-off and special events such as exhibitions, concerts (open-air events exert a special appeal), open days and sporting occasions.

Festivals also act as prodigiously successful mechanisms to attract visitors, always providing that, for each festival, its elements are distinctive and inviting, its location is accessible for its chosen audience(s), and that good marketing has publicised it adequately.

Temporary tourism features represent 'lightness' in a special sense; they are tourism butterflies, fluttering in to alight for a while, then leaving the picture. They are also 'light' because they are not rooted in a heavy, permanent physical structure, with the substantial staffing and managerial infrastructure that on-going attractions require for operation and maintenance.

Pilgrimage and serendipitous discovery

Many of the options discussed in this chapter have especial relevance and appeal in the context of Dynamic Tourism. Simplicity, abstraction and spirituality are central to many of these concepts. It can be said that the style of Dynamic Tourism stands in contrast with the early age of tourism, in the same way that British mid-to-late Victorian narrative poetry contrasts with seventeenth century Japanese haiku. One style is literal, wordy and heavy, while the other is characterised by reduction, measure and suggestion.

The leading haiku exponent, Basho, was also a journeyer, a foot traveller, who was, Lesley Downer (1990: 2) describes, 'like a priest, he owned nothing'. Through the personage of Basho, the notion of haiku is linked to travel and pilgrimage. Downer herself made a modern pilgrimage in Basho's footsteps, and the style of this trip, with the description given of it by Downer, communicates much of the characteristic essence of Dynamic Tourism:

> I had dawdled a little along the way. I stayed with friends in the grand old castle town of Kanazawa, cycling around the modern city centre and down the alleys of old samurai mansions and admiring the vast endless fish markets. Then I went to Ataka and wandered the wind-swept sand dunes, through the groves of spindly pines where the old barrier used to be ... And, like Basho and Sora [Basho's disciples], I stopped to take the waters at Yamakana spa. (ibid. p. 262)

In this passage, we can see the inclusive nature of Downer's travel, a necessary dimension emphasised in the theory of Dynamic Tourism. When travel is inclusive in its scope, more potential features are seen as worth visiting, and may be developed as part of the tourism repertoire and compendium.

Pilgrimage is a separate concept, where all experiences and discrete elements of the voyage gather together to form the dynamic. Features en route to an objective or a destination have an input role. Applied to tourism, pilgrimages would be interpreted as those instances where many diverse elements, rather than a single particular one, are regarded as an attraction, with each piece having a role of serendipity and potential significance. A dimension of spirituality is crucial to usual interpretations of pilgrimage, and may transfer well to the pilgrimages of tourism. Our awareness is commonly sharpened by new experiences, as tourism typifies and, with the intensification and range of newness of features, it would be unusual for no feeling of spirituality to be brought forth in the tourist from some facet or another. This, of course, draws in a key prevailing element of Dynamic Tourism – to extend places and reasons for going somewhere.

Simple, and without long permanence

Simplicity, as we have remarked, is in the very nature of Dynamic Tourism. To value simplicity is to pave the way for more additions to the usual repertoire of tourism products. For example, in the case of visits to houses and gardens, the habitual agenda would seek out only great mansions and ornate estate gardens. When simplicity is considered as an attraction, visits can be projected to see allotment huts, simple cabins and gardens maintained for utilitarian purposes in everyday lives. An illustration of this concept was given during the Cultural Capital year in Stockholm, where modern versions of allotment huts were shown. In line with the Swedish idea of a summer cabin, the 'allotment garden cottage' exhibit by Heikkinen and Komonen of Finland allowed sleeping and sunbathing space on a canvas-walled upper story, open to the sky. This exhibit, with its focus on simplicity, serendipity and transience displayed several core concepts of Dynamic Tourism.

A display of many corresponding characteristics was to be seen at the Northumberland country house and garden, Belsay Hall, in Northumberland in 2000. This was the 'Sitooteries' presentation of the simple entity of a place to 'sit out in' – interpreted at Belsay Hall as a garden retreat or other identified location of contemplation and refuge. Exhibits were of modern design, of great amusement capacity, and each individual portrayal showed varying degrees of complexity of aspect. The display was of manifest delight and charm to its visitors, of all ages.

Pleasures for all ages in a garden with temporary Sitooteries: Belsay Hall, Northumberland, England

Free and Public Space

As the world's population becomes increasingly urban, living in ever-greater concentration, simple space becomes an increasingly desirable commodity. The simple space possible in countryside holidays is an antidote to the inevitable congestion of the urban environment. To attract tourists to spend time *in* the city, it will be necessary to develop and rediscover more publicly available space. We have already identified parks as a key resource for urban space, and city squares, piazzas and other open and accessible urban spaces fall in the same category. To pursue both these town and country space trajectories, is to find more features to be used for tourism.

In the case of isolated areas that wish to court tourists, their very isolation may be an attractive feature; but tourism promoters must be aware that these destinations can suffer the deficiency of not being distinctive (Boniface *et al.*, 2000). The attractive element in one isolated site, unless it is unique, may be similar to attractions offered by direct competitors. Each attraction needs 'customising' or augmentation to make it special. In both town and country, as we have said, it will be necessary to increase, regain or establish public rights of access to provide more areas for a range of uses, of which tourism is one important facet.

As Richard Rogers (1995: 18) discussed his master plan for the river port of Shanghai, in his 1995 Reith Lecture, he explained that its aim was 'to transform single-use roads into multi-use public space – vastly expanding the network of pedestrian-biased street, cycle paths, market places, avenues, and making possible a substantial central park'. In the components he describes, and the provisions to open a landscape to tourism use and development in order to deliver a wider range of usable options, Rogers was in accord with the Dynamic Tourism outlook. His later remark (Rogers, 1997: 165) on cities, 'We must build cities for flexibility and openness, working with and not against the inevitable process whereby cities are subject to constant change' also coincides with the ideas that Dynamic Tourism propounds.

New Arenas, Combinations, and Widened Perceptions

Dynamic Tourism assumes a sophistication and readiness on the part of all concerned to go beyond the precepts of early modern tourism, where sightseeing was presented as a central activity. As an inheritance from the Victorian era, in which that form of tourism was defined, the sights judged fit for seeing were limited to the pretty and the picturesque. Now that our point of view as tourists is broadening to include our interest in lifestyles and the different ways people go about doing things, we find a much greater appeal in many more things, giving a greater extent to our potential travel and holidays.

Often one tourism product combines several possible kinds of appeal, thus allowing some seemingly everyday, mundane destinations to offer so much scope in context. Chapter 2, for example, describes walking as a tourism option. Extremely popular with the British, walking offers the traveller a 'blank canvas' on which to sketch in a variety of activities and ever-changing options, thus delivering dynamic flexibility. As the walker sets out for a known goal, determined upon as desirable to the tourist, the route to that goal has its own capacity of flexibility. The path followed allows for the arrangement of its components to all unfrequented and unknown destinations to be included *en route* to the final goal (Moulin & Boniface: 1999).

Links and Connections

A Council of Europe initiative links the towns of the former Hanseatic League. In spite of this formal connection, most of these towns, such as Bremen and Hamburg, are not currently regarded as tourist destinations. As de Graaf (1997: 21) reports on how these and other European ports are connected by communication, later making a link 'to the colonial world', he opines that this network of communication may be seen, in context, 'as a form of migration, pilgrimage, travel and, therefore, of tourism'.

While de Graaf discusses Europe, both as a whole and in relation to tourism and the rest of the world, he further defines and vaunts linkage:

> Vast stretches of Europe's farmlands are becoming available, possibly for recreational facilities. A fishing industry that has evolved into 'big business' has repercussions for many (former) fishing harbours, which now hope to become hubs within an international network of marinas; ports of call welcoming an increasing number of tourists on pleasure cruises. The new international network of high-speed trains also deserves attention; few speculations have been made on the potential of this medium from the standpoint of tourism. Permanent cross-channel connections and vessels capable of reaching increasingly higher speeds are making centres of what were once peripheral areas. Such means of conveyance, including mass media, revolutionise – time and again – the *conditio sine qua non*, the accessibility of tourism. (ibid.)

An example of new extensions

Rotterdam is a port that has not belonged traditionally to the portfolio of tourism destinations, despite featuring in the Netherland's Randstad agglomeration. Amsterdam, Delft and the nearby Hague have all been more visible centres for visitors. Recently, however, Rotterdam has been in the vanguard, as it showcases cultural attractions such as the Netherlands Architecture Institute. With its slogan 'Rotterdam is many cities', the city has been seeking to grow from its natural disposition as a 'multi-ethnic' port 'to becoming a 'fully-fledged multicultural city, and from an international into a cosmopolitan city' (van Meggelen, 1999: 2). In 1999, it was looking to see 'What will it take for Rotterdam to be considered a city of entertainment and leisure in the next century?' (ibid. p. 3). Perhaps it need look no farther than to capitalise on the city's diversity. Along with Oporto, Rotterdam has been designated as a Cultural Capital of Europe in the year 2001. A short-term participant in the year's initiative will be Riga, the capital of the Baltic State of Latvia.

The Change of Outlook and New Encompassing View

As the kind of tourist that Dynamic Tourism suggests we are, and needing the context to enable us to travel, we require a wider diversity of travel venues. These new possibilities should add under-exploited dimensions, attuned to our contemporary tastes, such as simplicity, spirituality, sensual appeal, modernity and abstraction. As these possibilities are realised, they should be designed to define and occupy more public space.

To find these potential tourism products, we must look for options beyond the traditional domain of tourism to elements of our everyday life. We must prepare to accept substitute additions. 'Synthetic' or 'artificial' products that are special creations, not replicas of pre-existing attractions, are obviously very useful. Temporary and light features, such as we have discussed in this chapter, should be considered and drawn into the realm of tourism, especially as they place no heavy demands on resources. Under-exploited portions of the world that are eager to attract tourism should seek to match what they have with what contemporary tourists prefer. Possibilities for networking linkages should be explored, to help publicise new products and to learn to develop by association with attractions already in existence.

The essential overall action that we need is to develop a broader appreciation of options in what we consider tourism attractions. As more and different products are developed and promoted, tourists will find new interest in their travel choices. The general task of the provider is to encourage new perceptions and to develop new aspects of existing destinations, while also creating new attractions, in order to give travellers more diversity and greater choice of places to which they can get away and go.

Chapter 5

Meeting

Dimensions to Encounter

The influence of experience

Since so many people are very familiar with being tourists, we expect that our tourism will continually develop. This is what Dynamic Tourism suggests, that we require our tourism to deliver new and deeper facets and to provide change in bringing us more valuable and more immediately personal experiences. While we were learning the accomplishments of seasoned tourists, we required the care and protection of the industry. We still need those services, of course, but as children do when they have learned adequately well in childhood to be able to stand on their own two feet in adulthood. When we travel now, we are not hesitant, we are ready and eager to go out among local people and seek interest and stimulation from them, their ways of life and their ideas. This chapter explores the question of how these optimum, happy meetings can be engendered more frequently.

Changed procedure, though meeting is not an aim

Even though contemporary tourists share a growing propensity to move into the host culture and meet its people, the tourist 'ghetto' will always have allure. Security is a primary reason, with a special appeal for novice tourists. It will attract us all on those occasions during our travels when we simply want to relax, to 'switch off' and avoid challenge and stimulation. There just are times when we do not want any new encounters, except maybe with fellow holidaymakers who share our general background and circumstances. We choose the option of the holiday package tour because it offers ease; we shop for it and buy it as we would any other commodity.

Today's package for the dynamic tourist must be based upon price competitiveness. The supplier's bulk buying must translate into demonstrable price competition. Once this is established, the package tour must offer much more flexibility of choice than yesterday's standard. Many operators are beginning to realise this fact. They cater to our new dispositions by offering a basic framework, with a minimum number of options, then allow us to choose from a considerable range of personal add-ons, according to our individual agenda. One example of this genre is the airline 'buzz', which offers, 'low fares and optional extras'. Buzz's understanding

Flying in different style: queue to board a buzz flight,
Stansted Airport, England

of the contemporary consumer are demonstrated by its publicity comments: 'we treat you as an individual not as part of the herd' and 'everyone wants different things from an airline, at different times'.

Conflict and benefits in meeting

Making the choice to leave the tourist 'ghetto' and go out to meet local residents and experience their culture can bring risks of conflict (Robinson & Boniface, 1999). Advance information and sensitivity are necessary in these encounters, if we want to reap benefits rather than problems and disappointments. Knowledge is presented as a key feature to Dynamic Tourism. Forearmed with information, dynamic tourists enter the experience by choice and with an open spirit, ready and eager to find the satisfactions of immediate human encounter. This element of human meeting is among additional situations to be deployed in tourism, and used by suppliers in developing more tourism products. Presenting further options under the 'meeting' heading assists with 'going', as described in Chapter 4.

Planning, chance and variation

Many encounters between individuals in the tourism context are seren-dipitous. This element of chance is one of the defining characteristics of the Dynamic Tourism philosophy. The particular connection cannot really be planned for – we can only arrange for tourists to know the circumstances that are most likely to lead to meetings, and then make their way towards them and court them. This chapter describes varied ways of bringing people together. It attempts to identify:

- Events that take people to new areas and experiences, those occasions whose style is meeting.
- Locations defined as meeting places, where people gather together.

Nevertheless, we must accept from the outset that, since people need balance and contrast in their lives, no one is always in a 'meeting mode'. As a counterpoint to sociability, on occasion we will demand that our holiday provides us with private times and solitude. One activity that both allows us to remain solitary, and yet indulges a taste for moving among other people and observing them is *flânerie*. This concept will be discussed as an important feature of 'meeting', in which encounters with people allow for the individuals to remain detached, and anonymity is retained. Cities are its natural environment, since in large populations observers of people can remain unobserved themselves.

In providing the possibility for a person to move among throngs of people without communicating with them, *flânerie* allows for pleasure, yet permits a measure of self-protection. This is not always a consideration, perhaps, yet it is a valid concern for the tourist who sees engagement as potentially threatening – as indeed it can be in some situations. In certain circumstances, in spite of the tourist's best efforts at gaining the prior knowledge to address a potential peril, a shortfall in street wisdom can put the traveller at risk.

On the way

An important part of meeting during travel occurs during the journey, before arrival at the destination. Opportunities for tourists to encounter one another and form friendships have long been recognised, both by the public and by the providers of those holidays where travel is the central feature: cruises, for example, or travel by specially commissioned transport, such as coach tours. For many of these trips, 'meeting people' is the principal aim of the participants. Often there is the unspoken assumption that people of similar demographics and interests are to be met on the transportation.

Even on everyday public transport, one well-known feature of the process is talking to strangers. Perhaps less noticed and thus under-promoted is the degree to which a journey on a holiday offers a prime opportunity for bringing together tourists from a range of groups. These encounters can happen on standard and large-scale public transport in the destination country. They may include individual tourists or whole parties of travellers, perhaps even from different nations, mingling with each other and also with the local population, which shares the same means of transportation. The younger generation, of course, recognises this phenomenon and thinks highly of these spontaneous, chance encounters. Young people welcome meeting new people, and often travel with the specific intention of seeking out such meetings.

Indeed, a foot-loose and fancy-free backpacking trip is almost a rite of passage for the young of the globe. The 'tales of the road' they bring back are important travel products, perhaps the most highly prized of their souvenirs. The Inter-Rail trip around Europe is one recognised example of the youth travel genre. However, one major difference to be noted in contemporary tourism is that holidaymakers of all ages, not merely adolescents, wish to participate in similar activities and enter into their spirit. This is true, even when the participants' age and maturity may endow them with different perspectives and expectations than those of younger travellers. Older voyagers may seek and obtain a slightly different, though complementary, satisfaction and timbre of holiday experience.

When travellers arrive at their destination, what combination of tourism elements is likely to serve in bringing about encounters? In general, these elements are some of the most fluid, unfixed and least formal of tourism options; their existence depends upon a certain premise. This presupposition is that Dynamic Tourism demands, and regards us to be:

- Less oriented toward objects as pleasure sources.
- More able and willing to find satisfaction through experiences and activities than through things.

Today's tourists are travelling for activities that present personal challenge. They want to find out about other people and their lifestyles, not merely to look at famous sights and visit museums. The situation is of the tourist wanting to be discoverer, as Dynamic Tourism posits. Contemporary tourism is characterised by less passivity and stereotypical activity; travellers today are growing into the new style of tourism, seeing what it is like and learning how fulfilling it can be. When tourists accept the premise that they want to encounter and know about other people in the places they visit, that acceptance 'lets in' the concept of meeting others, along with the attendant circumstances of such meetings, as travel objectives in themselves. When they are felicitous, such contacts have always been known to enhance a trip. Often it is the people met on holidays who are its most important memories and lasting outcomes. By this style of effort, a travel encounter becomes a desirable item in its own right, to be seen as a core tourism product, not merely as a chance by-product of the trip.

Open minds

As tourists we are now sufficiently sophisticated not to be over-concerned with weightiness, 'worthiness' and status in what we do on holiday; this contemporary frame of mind is also conducive to meeting people as an aim of travel. For example, when so many of us are able to fly long distances, and to see the world's great museums and tourist attractions, and when so great a number of people have already engaged in both these types of activities, new options start to emerge and are recognised to be attractive. We 'allow' them to join the acceptable range of tourist options. We permit ourselves to indulge a different disposition, that countenance of being 'light'. As Cohen has remarked:

> If the culturally sanctioned mode of travel of the modern tourist has been that of the serious quest for authenticity, the mode of the post-modern tourist is that of playful search for enjoyment. (Cohen, 1995: 21)

For example, we can accept Islamic Cairo as a main focus of tourist attraction, as well as seeking out the traditional Egyptian destination of the Pyramids. This sector of Cairo is a World Heritage Site, as the Giza Pyramids on the outskirts of Cairo form part of a World Heritage Site. Islamic Cairo's main focus of attraction touristically is not, however, so much its history as that it offers the tourist a close encounter with the residents' everyday lifestyle.

In fact, when we consider the Egyptian Pyramids, we might be ready to consider being content to see a synthetic replica of the real thing, as a mechanism to help preserve the originals from the stress of heavy tourism (as was discussed in Chapter 4, under the rubric of creating substitutes and man-made attractions to provide more tourism destinations). Such replicas may be found as 'fun' items in hotels in Las Vegas or at whatever other convenient locations and occasions suit the public and providers of tourism. In Chapter 5, we consider the message that it is fine to be as, supposedly, frivolous as to plan overtly to travel to meet new people and to enjoy pleasurable events with them.

Travel risks and how to avoid them

We must acknowledge that being 'let loose' to meet people is a situation fraught with possible dangers. Indeed, this is perhaps part of the *frisson* of excitement that tourists find as they engage in risky behaviour. In such an instance, a visitor is in the comparatively weak position of being far from home territory and its familiar modes and resources. Some protection from this vulnerability may be offered through the choice of *flânerie*, rather than direct personal contact.

If people meeting for the first time in a travel context have different language and cultural backgrounds, the contact situation becomes potentially more dangerous, or at least more fear-inspiring. However, in treating us as adults, Dynamic Tourism, with its criterion that we should 'hold the reins' in seeking out information and making choices, offers us a reason to talk to people. If we are given only the option of having to interact on a person-to-person basis in order to find out what to do in a particular place or circumstance, then we are offered the important chance to engage in dynamic dialogue. We can enjoy the possibility of a discussion of discovery with another human being, rather than being limited to listening to audio-tape advice or reading books of guidance.

Gathering places for contact

Well, then, how do the tourists meet other people? What are the necessary circumstances to allow an encounter, whether tourists choose the option to make full engagement with others by talking to people seen on their travels, or decide to move among people and act as voyeurs or amateur anthropologists observing the behaviour of others around them? Essentially, the important entities we must seek out are the places, events or occasions where people *gather*.

At first glance, this proposition would seem to focus more on cities and urban activities. To an extent this is true. But, since the term 'gathering' can be relative, the village fête in a small rural community can be the occasion of bringing people together, as can a religious congregation, or a food or craft fair relating to the cycle of agriculture. Urban locations representing meeting places include markets, parks, sites of religious worship, shops and shopping malls, and hubs of mass transportation. A manifestation of such meeting places can be found in the 'half-way house' territories of suburbia and exurbia; the particular forms peculiar to these areas are country parks, out-of-town shopping centres and airports.

Of course, certain entities that gather large crowds – festivals and special events prominent among them – can be very successful in drawing people, no matter where they are held, whether the location is town or country. The key aspect for all these 'gathering places' is that they serve as public, or semi-public, spaces and occasions. Some special events, most music concerts for example, are to a certain extent somewhat different in that they are exclusive because payment is demanded for attendance. However, such occasions are frequently made freely accessible by the providers, when they are seen as serving social, political or publicity purposes, often when occupying outdoor, public space.

Museums and galleries represent a domain that can be open to public access or not, depending on whether entrance is free or an admission fee is charged. In many such sites, since revenue is brought in by the ancillary activities of shops, restaurants and cafés, visitors can gain entry to these facilities without crossing the pay perimeter. Altogether, the options that are described here allow for fluidity and flux of people coming and going. New actors are allowed admission to the scene, and thus the situation provides dynamism, freedom of manoeuvre, constant novelty, fresh stimulation and opportunity for dialogue.

Engendering encounters

Some sort of background structure is needed as a context for the kind of occasions where travellers can go to meet new people. This need for structure is constant, even if the specific activity is short-term and impermanent in type. The framework of circumstances that lead to meetings must be pre-ordained. In those circumstances where the public dimension enters a private space, we still meet with formalised situations. Examples are schemes where private homes accept visitors, who either pay to look over a historic or interesting house or are paying guests staying in overnight accommodation. So, here you have tourism entering a private space. Tourists and permanent residents are brought together in these ways, with a style of meeting that is otherwise unlikely to occur between these particular people. When property owners or tenants elect to act as amateur guides and sources of tourist information, they participate in a loosely created engagement. There are other, entirely impromptu, situations where local residents and tourists meet – situations that are more likely than not to occur in the environments described above as lending themselves to such encounters.

Finding Appropriate Meeting Spaces

Public space and quasi-public space

In order for people to meet, it is crucial to have an arena of public space available. This observation is true, whether the contact is a general one of being among people and watching them, or extends to actual conversation. The very definition of 'public space' can itself be dynamic and open to interpretation. One interesting quasi-public space is Exchange Square in London. This site, already been mentioned for its Zen feature (a rock-

Relaxing in quasi-public space with help of Zen: Exchange Square, London, England

strewn water cascade) is only publicly accessible from Mondays to Fridays in working hours. Reporting Taylor's views, Sharon Zukin (1996: 46) says:

> The stores, office buildings, and theaters of commercial culture have defined public space for an increasingly mobile public.

Zukin continues:

> In commercial spaces the public are simultaneously customers and viewers, spectators and workers, at leisure and on display.

These remarks remind us of the shifts that are occurring in contemporary culture, they talk about the ambivalences of which areas do and do not represent publicly accessible entities. The passage identifies how we draw new and different places into use as public areas as we ourselves move around more, and it also notes that these new places often are endowed with a money-making function.

We may draw the further inference that tourists, and the locals whose acquaintance they seek, might meet with difficulty in finding enough 'neutral' places in which to encounter each other. Moreover, there is the issue of meeting a widely varied group of people, rather than a group limited to the people who are able and willing to pay the price of admission. The passage also implies that, where public space is needed, somehow accommodation to public pressure will be found. Certainly tourism today has no shortage of public realms open to occupation. It must be noted, however, that there is an occasional small obstruction in the fact that some sites are only qualified as public space, because their owners and operators demand payment for the right to occupy them. This means that some places cannot represent the complete gamut of people; they are exclusive rather than inclusive. Part of our function as dynamic tourists will be the obligation to keep genuinely public spaces in circulation. We must constantly look for a formulation of new types of public spaces, or a reformulation of old types in a new guise.

Festival marketplaces

In describing the 'commercial reuse of historic buildings' in the 1970s and 80s, Zukin (1996: 47) says that 'this spatial redevelopment harnessed an old urban vernacular – wholesale food markets, ports, railroad terminals' and she comments how, 'Many of these projects resulted in festival marketplaces'. This portrayal picks up on several of the categories of traditional meeting place and gives an example of the dynamic approach demanded to retain these spaces in public use. New ways of deployment must be devised in the replacement of redundant styles.

Already, the ubiquitous festival marketplace has passed on as a concept, in spite of the fact that many representatives of the genre are popular and commercial successes, such as Baltimore's Inner Harbor and London's

Covent Garden marketplace. The festival marketplace model is a difficult one, because the categories of shops are not mixed from a sufficiently wide variety. The routine types of shops that local residents need in order to make a habit of shopping there are often not part of the mix. Moreover, these marketplaces are located in *old* industrial centres, often far removed from the contemporary centres of everyday life. They are either defined as products intended for tourists only, or else they eventually take on that definition and remain unfrequented by the local populace except during evening leisure hours. Thus, festival marketplaces do not offer very good containers for varied general encounters between different groups of people.

Shopping, and markets

If we are to meet, as we so often enjoy doing, we need to accept as more important holiday features those places and activities that are conducive to meeting other people. Page (1995: 62) adapts Jansen-Verbeke's viewpoint to the consideration of the phenomenon of urban tourism alone and lists its primary features. His list includes such people-related aspects as Folklore, Friendliness, Language, Liveliness ... Local Customs. However, his 'value system' classes aspects such as Markets and Shopping Facilities as secondary considerations. Tourists, by contrast, nowadays tend strongly to want to be in meeting situations, so it is important that all the meeting types of elements be recognised as tourism features. Included among these features should be shopping as an activity and markets as a venue.

The proposition that shopping on its own can motivate travel is an easy one to demonstrate. If goods available at a certain destination are either unobtainable at home or are less expensive at the holiday place, then travel is encouraged. More especially, the motivation to travel is provided when regional items are available at only one location, for example individual craft objects, or traditional crafts made of locally-special material(s) to a locally-special design. Shopping also offers the possibility of meeting local residents. In fact, in relation to this consideration, the goods for sale do not necessarily need to be 'special' in themselves: they can be very routine. After all, the strong element of attraction to tourists and residents alike is the context to congregate among people. The architect Rem Koolhaas (1999), leader of Harvard Design School research into shopping in the city, has said, 'shopping is arguably the last remaining public activity'.

Markets and shopping deliver varied venues and locations. They have the potential for exotic and interesting sights, and the people who inhabit them are essential to their existence and vibrant nature. Aspects and emphases vary according to the size and type of market, and to the surrounding culture. While shopping is a necessarily constant feature in all cultures, in some instances markets can disappear altogether when a place is too modern to have a traditional open market, or too small for one to be viable. Souks are a type of shopping institution that takes the market dimension a step further; they represent a whole area set apart, with shopping as the major activity, and the whole gamut of daily life occurring on the streets within their boundaries. Thanks to their concentration and distinctiveness, souks can represent quite a challenging, albeit enticing, scene for strangers to visit and encounter.

Walking, strolling and meandering; city and countryside

The general notion of walking through city streets fully widens the dimension of dynamic travel, especially perambulating in the manner of the *flâneur*, in whose eyes all activities and people of the street are potential entertainment. By implication, Mazlish (1994: 57) connects the concepts of *flânerie* and tourism, saying, 'a passing type of the mid-nineteenth century, the *flâneur* ends up as an intrepid explorer of the modern and its consequences'. Shields (1994: 63) portrays the *flâneur* as 'like a detective seeking clues who reads people's characters not only from the physiognomy of their faces but via a social physiognomy of the streets'.

Paris is perhaps the pre-eminent walking city, for its size and variety. Of course, this is also appropriate, since Paris is the original city for *flânerie*, in the formal sense. Department stores were born here, and the nineteenth century saw an abundance of galleries with enticing shops and restaurants to be patronised and visited. Shields (1994: 65) opines that '*flânerie* is more specific than strolling. It is a spatial practice of specific sites: the interior and exterior public spaces of the city'. Shields' definition of these spaces includes parks, sidewalks, squares and shopping arcades or malls. In his view, *flânerie* is required to be purposeful and his opinion is that *flânerie* is public and other-directed.

Now, as the dynamic tourists are, by definition, well equipped with information about vacation destinations, they are likely to walk about to see things with a definite purpose and reason in mind. However, this is not to suggest that all walking and rambling must be directed to a pre-determined objective. This is far from the case. Strolling and rambling about a destination site, to see what happens and make chance encounters, are just as important components of Dynamic Tourism as are purposeful perambulations and tours. In fact, such ramblings are essential to the full realisation of the objective of meeting people, since a complete plan might entirely omit unknown elements in advance.

In the nineteenth century, it was daring for a woman to be a *flâneuse*. The French novelist George Sand, companion to the Polish composer Chopin, was notable as avant-garde in the activity, as well as in her personal approach to it. Of course, the general practice of walking around looking at things as a tourist predates *flânerie*. Remember the haiku poet Basho, mentioned in Chapter 3. In the seventeenth century, he explored the streets of the city of Kanazawa and inspected its fish markets. Today's tourist strolling the streets of Kanazawa can find, happily, that the city still boasts a fine fish market.

Let us reiterate: walking forms a key feature of Dynamic Tourism. Roles in the varying scripts of meeting and encounter, walking and watching, can shift continually between the tourist and resident, according to time, place and circumstances, but in any case, there is significant fascination. The city of Barcelona enjoys a key walking feature in its Ramblas route. Here, it is predominantly local residents who do most of the promenading. In many Mediterranean and Latin towns, the evening *passeggiata* is a prime feature, and one in which tourists can sit in sidewalk cafés and observe the community passing by.

City parks, with the many attractions they include, serve as appealing places for a stroll. They are natural 'points of people focus', offering chances for all varieties of people to meet and chat, as well as to have spontaneous conversations about the passing show happening directly in front of them. For example, there was a pretty merry-go-round in the Jardin des Plantes in Paris, unusual for its old-fashioned charm, and still more appealing for the obvious enjoyment of the children who rode on it. It was an attraction that drew observers, and among the spectator group were the parents of the merry-go-rounders.

Countryside walks offer a smaller likelihood of human encounter, but potential opportunities to view interesting features of the landscape and

Impromptu sight and encounter: merry-go-round,
Paris, France

rural activities are fully as abundant as the necessarily different urban attractions.

Eco-museums offer opportunities to bring tourists and local residents together by forming linkages to other, perhaps better-known, sites. In fact, planners for regeneration may use this potential for meetings between the different groups in their plans for social revivification. Devised routes of travel and circuits for tourists to follow can serve a similar role in integrating people who are isolated into wider contact and dialogue with travellers eager to learn about other ways of life.

Meeting: Toward the Objectives of Suppliers, Stakeholders and Hosts

It is often possible to notice a social or political subtext in provider actions that are ostensibly destined for visitors and aimed towards tourism. Suppliers and stakeholders, most particularly those in the public sector, may see particular kinds of encounters and understandings among people as necessities that must be engineered. People from among locals and residents might need to be brought together, and a tourism facility can be chosen as a mechanism and facilitator. Unplanned social benefits may occur through tourism activities, and it is plain to see that those in the 'meeting' category are likely to deliver these most.

One good example of this type of meeting dynamic is the Museum of New Zealand, Te Papa Tongarewa, opened in 1998. With its slogan, 'A place to stand for all New Zealanders' (undated Museum leaflet), it seems destined for a social mission of 'connecting peoples', intended to correct earlier omissions and imbalances in the outlook of the country's European colonisers. Te Papa is meant to 'reflect the contribution made by all cultures in our society and to explore the interaction between them'. New arrivals in the form of foreign tourists seem almost incidental observers in this central aim of the institution.

'Come join the dance' is an invitation to be involved, delivered by the Regional Tourist Board of the Auvergne in its holiday guide, *Auvergne: A Paradise of Space* (undated: 18). This guide vaunts the music festivals, country fairs and fêtes of the Auvergnat countryside, suggesting:

> At fairtime, go into the village cafés, the auberges, the farmhouses, and it will be pure theatre: people eat, drink, talk business, full of chat about the price of turkeys or fat cattle. But they also recount local anecdotes about butchers and dealers.

Describing the dynamic folklore the text describes:

> Auvergnats put into these fêtes all their good humour, their art of

living and their tradition. You should visit them, to see, to taste, to stroll around with your family. (ibid. p. 19)

The invitation seems offered to, not by, the local residents, who may themselves even be oblivious to it. Though doubtless meant with the best of intentions, and directed toward improving the economic and social fortunes of isolated and often poor communities, such publicity approaches to tourists can seem patronising to the communities that they attempt to market. Indeed, they may make local residents feel a bit like animals in a zoo, while tourists are encouraged as superior beings to come and watch them go about their quaint way of life.

A more equal and participatory attitude seemed to be present in Antwerp in Belgium, as the city mounted a huge, overall programme in 1999, celebrating the four-hundredth anniversary of the birth of its native son, the painter Anthony Van Dyck. To welcome visitors, 'The Helpful Citizens of Antwerp' provided, 'a competent citizen and *voluntary* [emphasis added] helping hand' (*Cultural Bulletin Antwerpen* summer 1999, programme until August 1999).

Consul-ANTS were also made available to organisations and groups, as an opportunity to achieve encounters during a visit; they were obtainable for a range of introductory talks, guidance and advice in Antwerp, or anywhere else that the organisations chose. One part of the festivities was 'Laboratory 99', a programme spread across workplaces in Antwerp, with visitors encouraged to join in activities as appropriate. The objective of the programme was to encourage mutual understanding and appreciation among representatives of different disciplines and, again, between these representatives and the general public. The whole city was regarded as a museum of 'urbanity and urbanisation', open to all comers, under the auspices of the 'Hetpaleis' scheme with Antwerpen Open. As this scheme was articulated, it was directed at 'the widest cross section of the public, from adults to children, from tourists to citizens, from the man in the street to the scholar, from rap enthusiast to little rascal!'

Another example of open, co-operative relations between a tourism site and its public, is the York Archaeological Trust in the United Kingdom. In the city of York, the effort to build stronger communication between a specialised profession and the general public (most especially children), led to the establishment of the Archaeological Resource Centre (ARC) by the York Archaeological Trust, a registered charity. The Trust has been a leader in the discussion of its professional activity, both as an obligation to the public and as a service to its own cause. The nature of the ARC is to be 'hands on', and it offers the opportunity to 'Meet real archaeologists' (ARC leaflet, undated).

Examples of People Meeting People

Australia: meeting on a trip

Australia has made backpacking a very strong element in its tourism. This activity, of course, connects strongly to the general ideas of Dynamic Tourism, especially the concept of travelling according to individual behest. It also lends itself to the particular notions of serendipitous encounters, meeting new people *en route*. In northern New South Wales in Australia is the 'hippie' region of Rainbow. Byron Bay, one of the region's population centres, is where 'Counter-culture is almost Establishment'. In the tourist brochure *Australia Unplugged: Escape and Discover Down Under*, when he describes the characteristics of Byron Bay and other towns in the same region and beyond in some other parts of Australia, Chris Tolhurst says that Australia is New Age. This dimension is characterised through: 'Earth worship, natural therapies, organic food, co-operation, anti-cynicism', and in saying, 'Insiders will tell you the movement is connected to the Dreamtime – Aboriginal Dreaming, though it's much broader than that: more like, a brotherhood of humanity' (Tolhurst, 1994: 13).

It is interesting that this piece has a passage devoted particularly to markets, presented as 'an easy way to tap into the ... counter-culture scene'.

Stimulation of new encounter: Paris

The whole joy of meeting new people, in tourism and at all other times, when there is a spirit of open-mindedness, interest and the positive hope of encountering somebody or something previously unknown, is that it can bring a sense of discovery and fresh sources of stimulation. For example, just around the corner from the Jardin des Plantes in Paris is an authentic mosque, with a tea garden, café and restaurant. Outsiders who step off the pavement of the busy European capital, through these portals, seem to visit an entrancing and fascinating 'other world' of exoticism.

Diverse and unusual encounter, ongoing from the past: London

The area of London to the immediate east north east of the City of London, the capital's financial centre, has long been a place where immigrants have congregated and settled. Its multi-ethnic identity gives it great appeal. Currently, the main residents of the area are Bangladeshi in origin. The heart of the sector is Brick Lane, and its many restaurants are a magnet for visitors. The area's street and market life display the full range of its ethnic diversity. In this district, North American immigrant Denis Severs chose to make a presentation to paying visitors at his eighteenth century house, 18 Folgate Street in Spitalfields, which he occupied in eighteenth century style. Though Mr. Severs is now dead, the show goes on, and paying guests are invited to meet the period by viewing rooms that appear as if just left by a family of Huguenot silk weavers. The visitor makes contact with another time by absorbing atmosphere and listening to sounds, as if the family had just left the room. The house is Mr Severs' creation. By bringing an artificial, heritage way of life to the house, he attempted to forge links with the eighteenth century, and visitors joined in trying to make the connection.

The chance to make direct and authentic connections with the area's inhabitants, both past and present, was provided among the Fringe events of the Spitalfields Festival in 1999. The organiser, Artangel, arranged for paid walks to certain special venues, introduced by writer and East End enthusiast Ian Sinclair. Among the places visited by means of a self-guided tour to see art works were the roof top of a private building, a bedroom of Brick Lane's Sheraz Hotel, and a disused Ashkenazi Jewish Synagogue that its current caretaker (the Spitalfields Centre) intends to convert to a museum of immigration.

One dimension of the Synagogue, which in its unaltered authenticity displays a more redolent atmosphere than anything mustered by Mr Severs, is in the sudden disappearance of an attic occupant, a mysterious Jew named David Rodinsky. Ian Sinclair and his co-author, Rachel Lichtenstein, who has put obsessive and dedicated research into delving into the secrets of Rodinsky's existence, have revealed Rodinsky's life and Jewish background in their book, *Rodinsky's Room* (Lichtenstein & Sinclair, 1999). Through the promotion of the tour and the book, as well as two other books written separately by Lichtenstein and Sinclair, this area of London has become better known. Moreover, the walk itself has delivered visitors to places, knowledge and encounters that they would otherwise have been very unlikely to find.

How the Role of Tourist Encourages the Process of Meeting

One important feature that should be stressed is how the role of being a visitor or tourist facilitates encounters and meetings with new people. Tourists are supposed to examine their surroundings, to stare about at things, to act in unexpected ways. They can do things that a local resident, with a long-term profile and image to maintain in the home area, would hesitate to attempt, for fear of risking the censure of fellows. Since tourists, in contrast, are generally ignorant of the new situations into which they enter, and are only temporarily staying in any given place, they are most frequently excused by their hosts for whatever they do.

Tourists, therefore, enjoy considerable license in their hosts' countries. While wearing the mantle of tourism and thus inhabiting foreign realms and making short-lived encounters in neutral places, we are given considerable leeway and flexibility of access. We can walk into other people's lives with no little impunity, as we often do while travelling on non-routine train or aeroplane journeys, where we confide in complete strangers and tell them quite intimate details of our lives without a second thought. We are, after all, speeding to destinations, with the security of knowing that this encounter is temporary, and that it is extremely unlikely that we will ever again meet up with this travelling confidant in the context of everyday existence. Such travel, in between our places of departure and arrival, is liberating for brief meetings.

The very means of transportation serves to create different types of connection as well. For example, relationships between the peoples of continental Europe and the inhabitants of the United Kingdom are surely facilitated and becoming more interactive with the ease and frequency of train travel through the Channel tunnel. This recent development is an addition to the communication and encounters that have, from the first beginnings of rail travel, been the accepted norms for Europe's mainland inhabitants and travellers.

Engendering Meetings Between Tourists and Hosts

The tourism industry

The essential dynamism and temporary nature of meetings between tourists and their hosts serve to heighten their impact, and can be part of the charm of attraction for all concerned. Essentially, it is not very difficult for the travel industry to create these encounters, or rather, to produce circumstances of a type that is very likely to produce them. The challenge for the industry is to arrange these encounters in such a way that they generate revenue, when so often they tend to be informal, random and individual, and their circumstances unstructured. In order to achieve benefits for the industry supplier, a lateral approach is required. The tourist industry itself must work in a strong context of meeting, by working in liaison and co-operation across a wide platform. Page (1995: 166) presents this context as imperative to successful urban tourism, and quotes Gunn's specification from 1988, that 'tourism planning should be pluralist ... [and] should be integrative'.

Regarding the meetings that it engineers, the tourism industry should sometimes look to collateral ways of deriving revenue from products that accompany encounters, rather than from the meetings themselves. This is especially true when the occasions for encounters occur in public or free space.

An onus upon the tourist

One dimension to be stressed about encounters with strangers while travelling, and this relates to Dynamic Tourism's precepts, is that the responsibility for the meeting rests essentially with the tourist. Travellers themselves must decide whether, when and how any meeting occurs. The type of tourist that Dynamic Tourism depicts would see any other way of arranging things as an insult. The role of the tourism industry is to understand this fact. It must respond to the requirement of providing situations and contexts for meetings, and must ensure that these occur in a context that meets its own needs and that in some way permit revenue to be earned, even if the methods of earning it are indirect.

One further aspect that must be mentioned is that encounters of discovery, for all ages and types of individual, should be supplied on demand. This imperative provides the opportunity for the industry to develop more products, as providers respond to the infinite variations of individual interests and needs among its clients.

Meeting Products to Spread the Load

There are two key dimensions in the process of conceptualising tourism products, which are directed toward encouraging encounters, and arranging the acceptance of those products within the tourism portfolio. In relation to this chapter's recommendation of an increase in the variety and amount of relevant types of products, meeting products render the visitor load as being spread and represent an extra range of product being brought in. By these efforts and outcomes described, Chapter 4's goal of encouraging travellers to get away on holiday thus, too, becomes facilitated.

Chapter 6

Spreading

Full Extension

A full range of products

Dynamic Tourism caters to travellers whose interests are wide and deep, who are motivated to see a broad range of sights. Dynamic tourists want and expect to see each point of interest in its entirety, rather than be contented with viewing the selected portion that the travel industry deems suitably glossy to be attractive, or sufficiently honed cosmetically and refined as a commodity to suit the tourist market.

Of course, there is a market for synthetic products such as theme parks, and also for simulations. Many of the products recommended in this book are models of these types. Artificial items may be exceptionally attractive in their way. Tourists like them when they match their current mood, and we must remember that these synthetic destinations relieve the pressure that mass tourism has placed on original, unique sites. However, recognising this one face of contemporary tourism does not diminish the importance of another view presented in this book; the tourist industry is insufficiently aware of the extent to which today's travellers are able and interested to see the *whole* of a real picture, rather than one that is edited.

Extending and changing methods of operation

One way for the tourist industry to expand is to court new markets. Since an assumption of the theory of Dynamic Tourism is that today's travelling audience is a sophisticated one, the corollary is to assume that dynamic tourists are already in the existing market, although their needs may not be very well recognised or met. This chapter will not focus much on tourism suppliers' lateral initiatives to find new customers. Instead it will concentrate on the discussion of extensions to products and methods of operations, in order to outline the concept of 'spreading' Dynamic Tourism by these means.

Methods of Spreading

Spreading can occur in three essential dimensions, as we have indicated. Extension can occur in product, process, audience; in space; and in time. It can be seen when the product immediately in use displays more, or all, of itself. It can happen through other items, aspects and subjects that were previously ignored in tourism now joining in to become part of a tourism product. Extension can occur when tourism moves to products similar to those already in use, but in new tourism locations. It can be represented by a producer's method of operation extending to embrace extra ways of approach, and as a reflection of a broadened attitude. Extension can occur when a tourist season at one familiar location lengthens and spreads into other seasons of the year. Finally, extension can occur when someone takes the initiative of bringing extra tourists or alternative markets to a product.

The Background of Gaps, and Its Causes

As they go about their business of providing to us, providers in the tourism industry should recognise that we now have potential access to full knowledge of circumstances, as was mentioned in Chapter 1. Customarily, the tourism industry has tended to offer a selection of certain types of products to their clients, assuming that these were the choices these particular clients would prefer.

In Chapter I, we saw Donald Horne's (1992: 201) discussion of subject lacunae in historical portrayals, along with his theme of the '*non*-representation' that occurs, 'when whole classes of people, even majorities, are ignored or may be present only in ways not related to their own views of themselves'. All this demonstrates how necessary it is to have extensions of information in textual forms, artefacts, or some other concrete presentation, in order to put before visitors the full spectrum of reality.

Some examples of incomplete presentations to consider might be:

- An old sugar estate in the Caribbean or an ante-bellum Louisiana tobacco plantation, where visitors are shown only the mansion of the master's family, but not the slave quarters.
- An English Victorian country house in which the enfilades of reception rooms and bedrooms of family and guests are open to visitors, but not the domestic 'engines' of kitchens and pantries below stairs, nor the servants' bedrooms in the attic.

Certain subjects and eras, with their artefacts, have been tacitly avoided, through the presenters' personal distaste or through a general lack of recognition on the part of providers that such subjects now have potential interest for visitors. Discomfort related to certain items could reflect distaste for themes and associations they call to mind, or an aesthetic judgement of their appeal. Industrial buildings, for example, were widely considered to be ugly. However, this judgement was made before today's changed perceptions, which recognise a special attraction in such monuments. There is onlooker appreciation of these monuments as canvases portraying technological advance and expertise. Also, a considerable interest has developed in the whole context that such buildings represent, with their story of mass working activity and its structure, as well as their relation to the wider social, political and economic scene of their times. Other categories that we have already cited (such as prehistoric sites and the abstract items associated with modern times and tastes) can be counted among potential tourist attractions which, when evaluated in the past, seemed to lack marketable elements.

Perhaps many excluded sites and subjects owe their neglect, at least in part, to the very natural human desire to show our 'best side' to strangers. Presenters may reveal this tendency. When acting as hosts to outsiders, we are conditioned to show maximum hospitality, and to present features in which we can feel pride. A dynamic tourist, however, wants to see more than showcases of the 'best' at a destination region; they want to see things in their entirety. The tastes and pre-dispositions of the dynamic tourist demonstrate an ability to enjoy and take an interest in the widest possible variety within the tourism product portfolio, as complete as possible in its representation and range.

Reasons for Suppliers to Extend

Tourism suppliers' core reasons for spreading by finding new products and seasons, are:

- To take pressure off over-visited sites, either during peak seasons or altogether.
- To bring tourism to places that want it *de nouveau* or in extended or greater amounts.
- Conservation.
- To bring social or economic benefit to disadvantaged areas.
- To equalise benefits which have been spasmodic or confined to short periods in the year.
- To make a statement about politics or status.

It should be added that spreading of these kinds is likely to be seen as an objective only by suppliers and stakeholders who are already able to think in strategic ways with an overall vision, rather than by those who think in a piecemeal and isolated way.

The Possibilities and Categories for Extension

More of the same; altering and adding to existing resources

While considering possibilities of 'spreading' we should first turn to those categories of destinations that are already perennial attractions. They are already well visited, even over visited, at least in certain sectors. Among this overall group are certain of the world's capital cities: Paris, London, Rome, Washington, Athens, Moscow, Tokyo, Vienna and Prague, along with other major cities such as New York and Sydney, and such historic cities as the European centres of Venice, Florence, Amsterdam, Siena, Toledo and Bergen. Other, slightly different destinations are Dubrovnik and Hong Kong. Despite their intrinsic 'blue chip' powers of attraction, these two cities have recently experienced problems with their flow of tourism – Dubrovnik with unpredictable waxing and waning, Hong Kong with a fall off in its traditional market.

All the places mentioned have their own distinctive niche in the tourism market, with their own recognisable kind of product. Most often, this product is perceived to be a heritage package, and so the city is identified by its historical monuments and buildings, and with their associations of power and leadership. Such cities' core focal points of tourist attention are traditional in nature, and remain essentially unvaried.

However, the dynamic tourist wants to see more than vaunted attractions and only a selection from a place. Such destinations as those we have listed can potentially offer more to tourists, if industry providers choose to do so. The reasons for making that choice might include relief from crowding at traditional sites such as fragile historic monuments and communities caught in a path of mass tourism. Another main reason for dispersal of tourism is to extend its benefits to other locations. These extensions are possible options because of changes in contemporary travellers' tastes, as we recognise in the theory of Dynamic Tourism.

Meanwhile, existing sites can also spread the impact of tourism through modifications in patterns of use. Such desired changes can perhaps be achieved by attractive shoulder-season and off-season price structures and other inducements, such as telling potential visitors about the enhanced levels of quality in the experiences that can be obtained by visiting during off-peak seasons.

Let us consider Paris as an example in our discussion of the concept of 'spreading'. It is clear that tourism occupies the historic city – the area that is conserved as a historic site and that represents the most overtly picturesque part of Paris. To the tourist's eye, this *is* Paris. However, since we want to cater to dynamic tourists, with their enveloping interests and freedom from fixations on seeing only the predetermined and traditionally appealing and attractive entities, other features of Paris can be brought into the tourism portfolio. As one instance of new destinations, there is the Villette Culture and Science Park; with its range of attractions and success with tourists and locals, it has demonstrated the potential to compete in this crowded market. The Parc André Citroën is also a successful new entry. Both these new offerings occupy former industrial sites and are viable attractions in areas outside the usual geographical boundaries of Parisian visits.

With the typical predilections of tourists having changed recently, if the marketers of tourism accept these changes and provide sufficient promotion for a fresh venue, the newly informed visitors ought to find their way to fresh experiences. An example of an attraction that should draw visitors beyond the Paris periphery is the Jardins Khan, which is a turn-of-the-nineteenth-century creation by Achille Duchens, a renowned landscape architect of that time. The gardens contain a series of garden scenes, expressing Albert Khan's concept of appreciation for global variety. Khan's ideal was world peace, through the means of international exchange. Within the original Jardins Khan, a new ultra-modern Japanese garden has been added as a complement to the existing Japanese-style components.

Still another non-traditional Parisian destination is the seminal Le Moult House on the Île St Germain, which lies in the Seine to the west of Paris at Issy-les-Moulineaux. Since it is a private dwelling, it can be viewed only from the outside, but it is worth the trip beyond the customary tourist zone to see this work of the contemporary designer Philippe Starck. With what useful social and economic effect could such appealing attractions as the Jardins Khan and the Le Moult House be promoted? To what effect could similar attractions be put, for example, if created in the multi-cultural milieu of the high-rises of Nanterre?

Both Washington DC and Moscow have small tourist areas. Both could obtain useful social and economic benefits for other parts of their urban area if they elected to extend their tourist regions. In both cities we may observe clusters of monumental attractions to be appealing to visitors in a traditional way. In Washington, the whole White House/Capitol Hill sector and the Mall with its major museums, when put together, form a visitor core whose farthest perimeter is defined by the Jefferson and Lincoln Monuments. The only traditional extension is Georgetown, with its canal, picturesque townhouses, restaurants, cafés and general air of affluence and power. In Moscow, Red Square and the Kremlin serve as the essential, central features for tourist visits. Together these two locations in the Russian capital form a World Heritage Site.

In Southwark, an area greatly in need of the benefits of tourism, London offers an interesting and ambitious example of extension in marketing. This formerly neglected sector lies very much within London's historic area; it encompasses a Roman bridgehead and approach, as well as medieval and Tudor waterfronts. Ostensibly, it should already be on the tourism map, yet it has hitherto been almost totally ignored. One explanation must be that Southwark, though once a place fit to house the Palace of the Bishop of Winchester, later declined into an artisan sector. With industry as its main focus in recent history, the neighbourhood has gone through a period of economic depression. It does not have much 'eye-appeal' in traditional tourist terms. Almost the only item to represent a picturesque exception to this rule and attract tourists by its historic charm is a working hostelry, the George Inn, which featured in Chaucer's *Canterbury Tales* and, much later, in Charles Dickens' *Little Dorrit*.

Location offers yet another reason for the former lack of tourists in the area. In counterpoint, consider Covent Garden, in a central position on London's north bank, this area hardly paused for breath in its transition from being a fruit, vegetable and flower market to becoming a major tourist destination. It should be said that, besides its advantages of location, Covent Garden also enjoys the benefit of more manifestly old streets that have considerable picturesque qualities.

A major effort is now under way to regenerate the neglected portion of the Borough of Southwark and turn it into a large tourism destination. One characteristic of this effort is that the local authority has commissioned architects to develop an overall master plan. The strategy incorporates a variety of features, many of them unusual or innovative. You might say they are working from the ground up, since the very pavements of the area are being brought into use as the base for innovative ways of signage to direct visitors to the local features and attractions.

A new Tourist Information Centre, designed for the purpose by architect Eric Parry, has opened near the railway station and Thames road-bridge crossing, in order to tell visitors exactly what is on offer and to help them find their bearings in this new tourist domain. Along with the traditional stock of printed leaflets, the Centre provides a display of area information in a dynamic format, to allow easy updating. Outside the Centre is the

Directing visitors on the streets:
Southwark, London, England

Southwark Needle, also designed by Parry, marking the place of Southwark's medieval gateway and serving as a landmark to welcome visitors and give them a visible point of orientation.

Food is one focus of presentation to visitors being developed in Southwark. The local authority is encouraging the use, on an occasional but regular basis, of the old Borough Market, once limited to wholesale fruits and vegetables, as London's Larder – a quality retail food market aimed at the general public. Trendy eateries and food shops are beginning to locate near the Market, and Vinopolis (a big visitor attraction discussed in Chapter 4) is settled here, along with its attendant food and drink outlets and restaurants. The character, style and atmosphere of this immediate area were, however, under threat in 2000 because of plans to improve a rail viaduct that passes overhead across the site.

Another natural theme for Southwark is Tudor life and Shakespearean theatre. The replica of the Globe Theatre is built here and serves as a focus for visitors, who come both to view the building itself, its fabric and form, with the Underglobe presentation, and to see it in use as a living theatre with the presentation of plays. This Globe replica is a private sector attraction. The nearby site of the original Globe is also displayed to anyone passing. Close by this original site, beneath a tower block, there is an imaginative sound-and-light show that presents in a broad brush, non-pedantic style, the archaeological site of the Rose, the first of the area's Tudor theatres.

Local development reaches out still further, however, with encouragement to practitioners of design and art, a theme carried through in the area's street furniture and visitor direction signs. The new Tate Modern art gallery is a major presence in its converted industrial site, the Bankside Power Station. In 1999, before the 2000 opening of this facility, the Tate delivered an outreach presentation of works by local artists in the form of the 'Bankside Browser', available in cyber form on the Internet and in 'hard copy' on request in a Southwark tower block. Also located in the general area are the Design Museum, the Jerwood Space art gallery and the Delfina Gallery. At the instigation of the Zandra Rhodes Foundation, these are being joined by the Fashion and Textile Museum.

Southwark, at the local Borough Council's initiative, wants to be overtly recognised as a new millennial tourist destination. Many of the visitor attractions are being packaged for presentation in a Millennium Mile, to be experienced as a riverside trail. In many ways, the development of Southwark is representative of a type of tourism destined to fit well with our current era and dispositions; many of Dynamic Tourism's preoccupations and requirements are reflected here.

Generally speaking, when we consider Southwark as an illustration of the concept of 'spreading' we note that a considerable part of the area will now serve as a major new tourism area for London. It will add an extra attraction to the customary tourist locations of London's centre and north bank. Its development in this role is to be helped, once the initial problem of over-much wobble has been righted and so the facility can actually be open, by the Millennium Bridge. The bridge was designed by Norman Foster and sculptor Anthony Caro to be an eco-friendly monument for pedestrian use. Here again, the attitudes encouraged by Dynamic Tourism are in evidence. Southwark, overall, represents spreading tourist territory in quite a large way.

The contributions of good access and transportation

Easy access is an enormous asset in developing potential 'spread' from a core tourist destination region to neighbouring areas that wish to become accepted into the visitor view. Good public transport is usually a key dimension in the access equation. Consider the example of Sydney, Australia, which is a port city, with a large, extensive harbour. Public ferries and special pleasure cruises (including the Harbour Cruise discussed in Chapter 4) ply the harbour waters and help to deliver tourists to farther-flung areas. The process of informing passengers is facilitated by State Transit Sydney Ferries, which provides a Souvenir Guide leaflet with descriptions of features of interest and attractions surrounding the harbour. The harbour trip itself helps arouse visitor awareness of the more distant excursion possibilities.

In Sydney itself, the Rocks area had gone into decline, but this oldest section of the urban landscape became the object of a Sydney Cove Redevelopment Authority programme. Now it is one of the city's main tourist areas, along with nearby Darling Harbour, an old port area developed during the 1980s. There is a Visitor Centre at the Rocks, where tourists receive a general orientation and encouragement to take a walking tour of the area. The National Trust also has a Centre in this area, and, to help tourist information 'A Story of Sydney' is delivered at yet another venue. The oldest house in Sydney shelters the New South Wales National Parks and Wildlife information point and bookshop. The Australian Wine Centre is housed in a rehabilitated storehouse. Australia's most famous textile design artist, Ken Done, who is credited with assisting in defining Australia for both residents and tourists alike (Powerhouse Museum, 1995: 46) has a gallery housed in an old warehouse in the Rocks.

The Rocks also offers lots of shopping opportunities. It is typical of such waterfront developments, which often occupy and refurbish decayed industrial portions of cities, and to which tourists are encouraged to come to enjoy the shopping and ambience, while helping to effect regeneration through the money they bring into the area. Proximity to Sydney's downtown has made it easy for The Rocks to develop as a tourist colony, as it was earlier a colony of settlers. Darling Harbour, with its Harbourside Festival

Market Place shopping, City Convention Centre and Australian National Maritime Museum, is a less well-located visitor venue. Access here is facilitated by an overhead monorail that follows a circular path, crossing water on a bridge that is otherwise reserved for pedestrians.

Examples of Spreading

Dispersing tourists and spreading the product: Venice

Venice is perhaps the pre-eminent tourist city. Access and transportation within its bounds involve boats on canals, and pedestrian traffic along streets that cross canals by footbridge. The implication is that, as such a much-visited city, no part of such a contained space as this city represents could fail to be in tourist occupation. The reality of Venice is somewhat different, with tourist zones being created in specific areas where there are major monuments, ferry stations and bridges. If Venice were to decide to disperse its concentrations of visitors and extend its tourism to less-frequented portions of the urban area, altering ferry stations and building new bridges would probably help do the job (Boniface, 1999: 150). The dynamic tourist, of course, would be expected to seek out workaday, contemporary Venice and points of interest off the beaten track of the city's famous, old and clichéd attractions.

Extending out from a tourist ghetto: Bergen

In the past, historic areas, and especially those that are overtly appealing by being picturesque, have represented the natural destination focus for visitors. With this emphasis and believed visitor preference, many cities and towns have developed ghettos of tourism inside them, with their other parts being largely untouched by visitors. The waterfront city of Bergen is essentially known in tourism terms for: its Bryggen historic warehouse area, which is a World Heritage Site and which has been prettified in the way many such places are as they lose their multi-functionality and working-class life; Greig's house in the suburbs; and the quick access it offers to nature and the countryside. Bergen's main shopping district, which offers little in the way of merchandise that could not be found in any similar shopping area in Norway or other Scandinavian countries, is a more minor feature of promotion. This leaves an open option to large pockets of the city of Bergen, most of it in effect, to promote their own features to tourists who are sated with either old historic quarters living off a cloak of picturesqueness or off nondescript shopping opportunities.

Different market, wider type of product, longer visitor stay: Hong Kong

Hong Kong has a developed role as a stopover on long air journeys, such as those between the UK and Australia. Shopping has always been another part of Hong Kong's appeal. The attractions of Hong Kong have been those of a frenetic, urban type, and have focused on Hong Kong Island and Kowloon, both of which have mighty population densities. This sheer concentration of population and activity has been a pre-eminent, distinguishing feature in visitor perception and experience for Hong Kong. In its post-colonial status, Hong Kong has met the experience of a waning response from one of its traditional tourism markets, British travellers. Since its new position as a part of the People's Republic of China has changed its character, Hong Kong is forced to find new ways to attract certain audiences that used to enjoy visiting a colonial milieu.

Hong Kong's traditional role as a stopover point, breaking a long flight *en route* to somewhere else, coupled with its power to overwhelm through frantic activity, has been a reason why the city has not been seen as a long-stay tourist destination. At the height of its popularity, with the loss of its colonial status impending, Hong Kong began the process of developing ways to extend its tourist appeal. The Hong Kong Tourist Association (HKTA) advocated a spread in visit length with the modest, but doubtless realistic, suggestion to 'Stay an Extra Day'. Also, a more diverse portfolio of attractions was promoted, in order to counter the visitor perception of Hong Kong as merely an exotic shopping destination where travellers might enjoy the fringe event of a meal at a fish restaurant in one of the villages of Aberdeen or Stanley on Hong Kong Island's south shore.

Visitor attention was newly drawn to cultural, historic and religious features, as well as to beaches, and above all to nature through the promotion of Country Parks and hiking trails. Victoria Peak is the ultimate residential location of Hong Kong Island and the site of an eye-catching shopping opportunity. Even here, the authorities provide the Victoria Peak Circuit, with its nature information boards. The Circuit was one of the Country Walks offered under the 'Stay an Extra Day' promotion. However, one may entertain the suspicion that this path serves more for executive jogging than for the discovery and enjoyment of tree and plant life through purposeful strolling.

In 1993 Governor Patten inaugurated a new type of heritage trail in Hong Kong, the Ping Shan, a co-operative venture based on an idea from the Antiquities Advisory Board. The governmental Architectural Services Department (Antiquities Section), the Lord Wilson Heritage Trust, the Royal Hong Kong Jockey Club and the Chinese clan of Tang all joined forces in this project. To show visitors the unexpected part of Hong Kong, on an extension beyond the customary tourism sites, the HKTA provided the 'Land Between Tour', which travelled through the New Territories and revealed Hong Kong's more rural and traditional areas. The tour visited a temple, a market, a fishing village and the border that still existed with

*Extending to bring a market into view,
and to bring a new opportunity to meet:
the New Territories, Hong Kong*

China, among other sites. This tour made a special effort to show more of northern Hong Kong to visitors. The opening of Chek Lap Kok, the new airport on Lantau Island, designed by Norman Foster and linked by a bridge to the main portion of Hong Kong, gave an immediate focus to bring visitors out in a westerly direction from the central, heavily-visited area.

Extension of perimeter and product: central Tyneside

The former Baltic flourmill in Gateshead, in the United Kingdom, is acting as a new type of focus in a new guise; for its fresh function as a visitor attraction, it is now to be known as BALTIC. Gateshead lies across the River Tyne from the regional 'capital' of Newcastle, and has suffered an insignificant, low-profile role in relation to its larger neighbour. Although neither city has been a major site for tourism to date, Newcastle has been the greater draw, if only to visitors passing through it as a gateway city, since it holds Tyneside's main railway station and has an airport not far to its north.

Most tourists travel to Newcastle either with their attention focused on the coastal castles and beaches further north or with eyes looking westwards to the visible stretches of Hadrian's Wall. Funded by a large lottery provision through the Millennium Commission, the International Centre for Life, through its Life Interactive World visitor attraction, now offers a reason to stay in Newcastle. Although the perception is neither public nor explicit, Gateshead, in its turn, must hope and expect that by providing (also with huge lottery funding help) the BALTIC as a major international attraction, it will draw visitors and their associated economic and social benefits to its needy area. BALTIC, in the words of its Swedish Director Sune Nordgren, is to be an 'art factory' (*Baltic* No. 1) to hold temporary exhibitions. A shop, café and panoramic restaurant are planned as part of its complex. Alongside BALTIC, and contributing towards the building of a larger overall object of enticement, will be a Regional Music Centre designed by Norman Foster (also with a large grant of lottery funding).

In common with Southwark, Gateshead, too, is to have a new bridge to bring visitors south across its river from a more frequented north bank. This is an important factor in achieving 'spread' and associated regeneration. This Millennium Bridge, which does not open until 2001, is intended to increase the link between the two Tyneside urban settlements; it will serve pedestrian and cycle traffic only. When BALTIC began its pre-opening publicity, it began the process of luring people to its activity and its eventual site by 'invading enemy territory' and holding exhibitions in the Hatton Gallery of Newcastle University. Outreach will always be a key feature of BALTIC. One more notable, telling item, is the fact that Newcastle City itself is a participant in the overall Gateshead Quays project, of which BALTIC is a focus. Certainly the co-operative relationship between these Tyneside cities will strengthen the tourist market for both. Over the years, interesting comparison will be able to be made between the tourism effectiveness of the two, in many ways alike, projects: the Tate Modern in an old industrial building conversion on Southwark's waterfront, and BALTIC similarly housed on the waterside at Gateshead.

As Tourists Change, New Items Move into the Market

Formerly unfrequented tourist areas

One objective of spreading is to draw visitors from heavily populated parts of a destination region to areas less frequented by tourists. There is an equal need in rural areas, as in cities and towns, to extend tourism. In the instance of Hong Kong, the plan for spreading is directed from urban to rural regions, while a second aspect of the plan allows for spreading the *type* of tourist activity. As the world population shifts to cities and urban areas, country populations may be left behind. The economic and social benefits that tourism delivers can be very much in demand among cities and urban conglomerates, but may often be even more crucially needed outside them, in the countryside and across more widespread geographical areas.

The environs of cities, especially when the cities they relate to are in post-industrial decline, can be as needy of tourism as rural areas are, but they often possess less obvious benefits and contrasts to offer visitors than country destinations do. It would be difficult for such areas to deliver traditional tourism, with its requirements of standard attractions and accepted transport facilities. Problems can be fewer when the target tourist market is dynamic tourists, who may find a wider range of attractions, activities and modes of transportation to be desirable and acceptable.

One of the premises we have adopted is that, as everyday life becomes more complex and mechanised, consumers want a correspondingly greater portion of tourist time to represent a contrasting element. England, even though perceived as an over-marketed tourist country, still has areas that remain unvisited. For example, take the castle town of Conisbrough. Here we may observe a brownfield site that is being transformed to a green one. The general area has many current and former industrial sites and, in consequence, a large nearby residential population. It is here that a new kind of attraction, the Earth Centre, alluded to in Chapter 3, is situated. The Earth Centre's style and subject cater strongly to precisely the kind of tourist that this book calls 'dynamic' – to such an extent that we will discuss it in greater depth in Chapter 7.

Another striking example of a site that conforms even more closely to the Dynamic Tourism style is to be found in England's industrial East Midlands, where a pig farm and its accompanying wasteland have been transformed by a Japanese Buddhist monk into a Japanese garden. The transformation, which involved use of some recycled material, has turned the site into a tourist attraction. Pureland Japanese Gardens are now well established as a free entry site. There are plans to provide a café in an old barn to generate income and improve visitor services (Taylor, 1999: 3) The Earth Centre and Pureland Japanese Gardens both extend tourist activity into areas where it was absent, turning industrial and agricultural waste-land into viable, income-producing attractions.

West of Clermont-Ferrand in the Auvergne in France, the European Centre of Volcanism extends tourism from existing visitor attractions in a rural area, as it builds on, and out from, its region's natural features of volcanic puys. Here visitors may experience a simulated journey to the centre of the Earth. We have already discussed other attempts to market tourism in this region, which describes itself as 'dynamic' in the undated *Centre Européen du Volcanisme* leaflet. Besides welcoming tourists, the complex serves as a conference and documentation centre on Earth sciences. There is a restaurant in the grounds, and natural green areas are part of the overall design.

Generating and receiving

It may often seem that the world is over-run by tourists, but large parts of the globe are neither destinations nor jumping-off points for pleasure trav-ellers. Those sites that do receive visitors, may do so only for pockets of their whole circumference and so have much spare capacity for extension. There are vast countries such as Canada and Australia where populations are thinly spread in many parts; they could receive many more visitors across a much greater part of their area. Of course, it is necessary for the host countries to want an increase in visitors, and for visitors to find these isolated, scorching or frigid territories appealing. Both Canada and Australia, however, do indeed already serve a great number of tourists.

The countries of the former Soviet bloc also represent, overall, a massively under-filled capacity as destinations for tourism. In the main, these nations are as yet unvisited by international tourists, notwith-standing the converse for the 'hot spots' of Moscow (a portion only, as has been said), St Petersburg, Prague and Dubrovnik. The more southern portions of the huge general area are already operating in the tourism domain and as sunshine resorts – those of the Black Sea and the coast of Croatia are examples. By implication, the northern and rural sectors of this vast area could extend and develop a tourism industry, should the nations involved wish to do so.

The Baltic Republic States of Estonia, Latvia and Lithuania are emerging rapidly into the daylight as tourism venues. They have coasts close to Western markets, and their capital cities are of intrinsic interest. These countries are being quick to accommodate tourism, and standards are being raised to the level necessary to attract international visitors. The casual sighting in London of a tour bus from Riga reveals that visits between Latvia and England are two-way, and so likewise bilateral are the spreads of tourism represented. After all, Riga is to join Oporto and Rotterdam for a short while in 2001 as a Cultural Capital of Europe, a matter of significance in relation to its possibilities of developing and extending its tourism presence.

The WTO opines that the Baltic Sea destinations are now in a position to grow as tourism venues, since they now project a stronger image and have 'improvement in the transport infrastructure' and 'advantages ... [of] safety and political stability'. Eco-tourism is one of the aspects that the WTO suggests as an area of concentration for the Baltic Sea resorts. The WTO's Chief of Quality Tourism Development, Hendryk Handzuh, articulates the view that 'the Baltic Sea region can be the place to promote *true ecotourism*'. While commenting on the extent to which 'Hanseatic and Viking itineraries have already made inroads in many people's minds' (WTO, 1998 press release), he also mentions another focus point, cultural tourism.

Scandinavia also has room for more tourists, if the region's countries wish to attract them. The strong environmental concerns of the area could dispose them against inviting additional large numbers of visitors to their territories. A natural extension of travel in Scandinavia would be tourism to Greenland, a possibility promoted by the book and film, *Miss Smilla's Feeling for Snow*. As the less orthodox tastes we have noted and called 'Dynamic Tourism' emerge, and with a less pronounced fixation among travellers that holidays equate to time spent in sure, sunny weather, northern Europe has the potential for more tourism to extend into its area. The countries of northern Europe possess the advantages of maturity and experience that developing countries, inevitably, lack. Thus, even if these northern European areas have not been very active marketers of tourism in the past, they have a general background of knowledge and expertise that can serve as a basis for moving into tourism and running the activity successfully and responsibly.

Adding to an Existing Site to Appeal to More Tastes

Of course, tourism extensions need not necessarily cross the threshold of an existing site in order to cater to wider tastes and bring in new dimensions. If there is a preference or necessity to limit a tourism area to its existing physical boundaries, it is more than possible to do just that. A good example of such extension within space constraints is the Meditation Hall, which was introduced recently on the UNESCO campus in Paris, alongside the Noguchi Garden (discussed in Chapter 5). Designed by Tadao Ando, the Meditation Hall is architecturally abstract and simple, and defined as accessible, since its portals are doorless. The drum-shaped Hall is surrounded by another part of Ando's concept, a shallow, terraced stream of running water, crossed by ramps that lead to the portals at each side of the building. Neutral of stylistic rhetoric and elemental in its appeal, the Meditation Hall presents a haven of attraction and contemplation, like a cool cave girded by running water.

Brand Extension

One way of spreading tourism that has been used with success is extending the 'brand'. For example, Bilbao now has a Guggenheim attraction, an art gallery that has enjoyed immediate success as a catalyst in bringing this Spanish post-industrial waterfront city and port in amongst tourism destinations. In the United Kingdom, the Tate brand has been similarly deployed. It is now being used in Southwark, across the river from its Pimlico home base, as we have already described. In fact, the Tate brand name was first taken to Liverpool, as part of a festival marketplace in Albert Dock. From there it went to the small town of St Ives, on the north coast of Cornwall.

Encouraged by the introduction of railway connections, St Ives has long acted as a tourism destination, mainly as a beach resort. Many artists have made their home in the area, and this artistic dimension has contributed an additional charm to its appeal. The decline in the traditional 'bucket and spade' market of St Ives has left room for a spread of greater attention to art in St Ives' tourism personality, with tourists flocking to the gallery that marks the extension of the Tate's brand presence. The Tate name is bringing a new and affluent market to St Ives. Since its motive for visiting the area is largely independent of the weather, this group arrives throughout the year, thus delivering both a time spread and a volume increase in tourism – to the benefit of the town.

An important feature of this extension is that, in relating to art, it is both connected to and rooted in the town's prior existence. A 'spread' has been selected that is characterised by credibility and authenticity. Since the whole of St Ives had a standing context as 'arty', there was an overall framework of pre-existing support for this spread and development. The depth and critical mass of this pre-existing support, meant that visitors are provided with more added value and increased interest than if the Tate St Ives were standing in isolation. The Tate St Ives, in its excellence, and style of product and presentation, offers a shining example of what shifts in standard other tourist providers in the town must make if they decide to cater to the defined, discerning, and potentially high-spending audience now drawn to their midst. Among those new standards are:

- Style of appearance and high quality of maintenance. The Gallery offers in its building (designed by architects Evans and Shalev) certain 'pictures' of St Ives through strategically-placed windows that add the local scenery to the core product of paintings and sculpture.
- Quality overall, and add-on services bringing benefit. The Gallery has a sophisticated café-restaurant and maintains a bookstore of sufficient quality to be an attraction in its own right to a specialised, cultivated market.

In terms of Dynamic Tourism, perhaps the special invitation of St Ives is that it offers an unusual and varied combination and spread of attractions. This is that the features of a traditional seaside town are now elided to an art gallery of international standing, along with certain fringe elements of similar vein that have been encouraged to emerge.

Spreads in Time

Extent and depth of history: Lyon

Spreading can occur in a single place, with the aim of including more, or even of being all-encompassing. The French city of Lyon has maintained a low profile as a tourist destination, perhaps being best known for its gastronomy, and for its place as France's second city, rather than for any particular dimensions of architecture or visitor features. Promotion of Lyon as a destination does not single out any one element. This is indicated by its 1998 designation as a World Heritage Site, the distinction being accorded for its overall historic area – for the elements of all eras in Lyon, rather than any particular one. This novel, inclusive attitude delivers a product well suited to Dynamic Tourism. Lyon's product offering represents spread in time.

Widened view of the tourist day: Antwerp

Antwerp demonstrates another way to practice temporal extension of the tourism product. Here, tourists are invited to view at sunrise, and by bicycle, those city elements needing to be at work at the time. Departures are at 4am. This is in addition to Antwerp's offer of good and unusual sites for the rest of the tourist's day.

Extensions to tourism seasons and their points of emphasis

Another type of time extension, as St Ives reveals, can be in visiting seasons. So many destinations are constrained, or believe they are, to a seasonal limit in their appeal, tied to the tastes and travel possibilities of their clients. One classic way to alter this situation, as in achieving any shift of change in product use, has been to offer inducements, especially price reductions. Glance at a travel brochure, and you will recognise the management mechanisms that underlie cost ranges. The most expensive holidays tend to be at the times of school holidays, Christmas and Easter. At the other extreme, in terms of price, are the off-peak periods, when appeal is at its lowest point. Any significant climate variations that affect visitor appeal for a venue or trip will contribute to setting prices, in order to encourage visitors to spread out their times of arrival.

Strasbourg is a good demonstration of how an 'out of season' can be manipulated to produce an 'in-season'. Strasbourg has a Christmas Market that runs from the end of November to the close of December. By providing

Night activity, and 'out of season': Christmas Market, Strasbourg, France

additional associated activities, and promoting itself as the 'Christmas Capital', the city has become a visitor-filled venue during the period generally regarded as tourism's lowest.

A related way to extend tourism in the present context is to court new markets. The tastes of dynamic tourists, who want to see what places are like 'in the round', can extend to wanting to see sites through the fullness of time as well. These tourists can be encouraged into visiting a resort out of season, attending a summer festival's location in the depths of winter, even occupying a summer home or cabin during the year's coldest months. The dynamic tourist would not automatically restrict visits to Paris to the month of April, as is the cliché, for example.

The disposition of dynamic tourists to be 'spreaders' is helped by so many of their kind being included among the 'grey', early-retired, single, young, self-employed and unconventional categories. These groups are relatively foot loose and fancy free and may not be tied to the needs of employers or others. They may be in a flexible situation that allows them to go on impromptu holidays. They may be able to go on a trip at not accepted or governed times. This could have an impact for an area being over-visited at one particular time. Take the example of the Mediterranean, a leader among areas that are over-visited seasonally. The situation is encouraged by the disposition of many Europeans to holiday in this region, and to do so in the short period of high summer. The company Explore, which offers 'small group exploratory holidays' because 'small groups display smaller footprints' (undated *Explore* leaflet), and which caters to the dynamic tourist type, has noticed a change in trends. The trend is now to off-peak travel to the 'southern European Countries bordering the Mediterranean and Aegean'. Explore has promoted this trend further, telling its market in the United Kingdom that the off-season 'can ... be the best time to discover the *real* flavour of popular destinations closer to home' (*Explore* Newsletter, March 1996).

Extensions to Originals

As substitutes and extras

We have already addressed the question of authenticity in an earlier chapter; it is an issue that is frequently raised. We recognise the fact that replicas can serve as substitutes that help conserve fragile originals. Such replicas are extensions of the original attraction. On the subject of taste and tendencies toward this particular types of spreading, the question has been raised whether tourists of the new persuasion care more or less for authentic items, and whether authenticity always really matters very much to the tourist of today.

Furthermore, the debate can centre upon the definition of authenticity for the tourist, and upon how the interpretation of authenticity of both provider and tourist may be conditioned by culture. The Shakespeare Country Park in Maruyama, Japan, is an extension of Stratford upon Avon, a heavily-visited small town in England that is famous for its links to the playwright Shakespeare. Japan has produced an audience of many Shakespeare devotees, and hence a market for a local Stratford upon Avon replica. In the Shakespeare Country Park, the buildings reproduce in English oak the style of the Stratford upon Avon originals, although they incorporate twentieth-century compromises. There are exclusions and improvisations. Pearman quotes the established architect, Julian Bicknell (1997: 8), who is responsible for the enterprise, as saying, 'the Japanese see things differently ... their attitude is not wrong – just the expression of a different cultural outlook, one that places metaphor higher than literal-mindedness'. Bicknell sees the Park as representing 'an educational and cultural resource'.

The Available Ways to Spread

As we can see in many of the examples discussed above, the tourists to whom Dynamic Tourism is set to cater are the kind who want to extend their tourism into new dimensions. Because of this new style of contemporary tourism, it becomes possible to spread to new types of products and new ways of operation. These extensions can be to existing products, they can be the same type of product but developed in new locations, they can be new kinds of products, they can be wider seasons or different times, and they can also be moves into different markets.

Though this issue has not yet been addressed in this book, spread can also be an extension in the process of making products available. Internet sales are an obvious example of that kind of spread. The essence of the spread addressed in this chapter, however, is the disposition of dynamic tourists to know a whole dimension and position in the tourism arena and their wish to encounter the reality of each tourism product.

Chapter 7

Moving

How to Move and What to Move to

Dynamic Tourism's credo is that tourists have already changed and that they are continuing to evolve. It follows, therefore, that the tourist industry must provide still more than a simple increase in existing products. Fresh and different items are required for dynamic tourists. They want the opportunity to experience the full range of what the world has to offer, not simply a selection of tourism products, fashioned expressly for tourism and chosen for them by the industry. Products meant to appeal to the dynamic tourist must be characterised as much as possible by flexibility, to suit the changing predilections of this market. This chapter will discuss what these 'new' products might be, as well as the features they should display. When the word 'new' is used here to describe tourist attractions, the item described may be literally new, or may instead be an existing product presented in a new format among the field of tourism offerings.

New products can be found by:

- Identifying products as belonging to the tourism portfolio, when they have not been included there in the past.
- Revisiting and reinterpreting existing products to give them a new face or to present their original features with a new, revived, appeal.

While this chapter will discuss both of these categories, the first will carry our principal emphasis, since Chapter 8 focuses on the replenishment and refurbishment of old attractions. While recommending products that might encourage visits to sites yet to be visited by the travelling public, Chapter 4 identified a certain number of new items.

What to move from, and why

In order to distinguish present and future novelties in tourism, it is helpful, indeed, necessary, to begin by identifying those old products already in existence. Such identification helps to define the difference between old and new, and immediately implies the sort of general 'moving' that providers must carry out if they are to bring new products into circulation and put them before the public.

In order to deliberately create more attractions, so that a greater travelling public may reach destinations in comfort – to itself, resident communities and to sites and environments – suppliers must carefully identify the general elements of tourist appeal. Such identification is important; in order to meet its own objectives and those of the greater society, the tourism industry cannot rely on mere responsiveness to groundswells of existing tourism demand and public opinion. It must anticipate them. Members of the industry must know audience impulses well enough to identify kernels of interest and build on them in order to meet market demand with the right products. One occasional role for tourism providers will be to *generate* interest from a basis of knowledge of the public's potential and latent interests, and the predisposition of the market to accept a proposed product. Thus, a budding tourist sensibility toward a new attraction may call for development, rather than being left to grow without encouragement.

Traditional tourism products are:

- beaches – sun, sand and sea;
- ski resorts;
- theme parks;
- museums:
- scenic attractions: of those that are man-made, mainly historic monuments; among natural features, principally mountains and lakes.

The oldest expected roles for tourists in relation to these are those of being passive, relaxed, and looking at sights. Only the relatively recent introduction of skiing as holiday recreation demonstrates a leaning towards showing more energy on holiday. Dynamic Tourism tends to favour active pursuits. However, it should be emphasised that 'old style' features are not expected to vanish altogether. On the contrary, these traditional characteristics and motivations will persist, but will be seen *alongside* newer propensities and trends. The situation simply calls for a move to characterise a wider group of products as suitable for the tourism market. Companies will maintain older offerings. Some of these may need to be revised and adapted, but these traditional products will be accompanied by a large number of new additions.

The Type of Items to Move To

Key items among novel tourism products that are presented in this study as part of the essence of Dynamic Tourism are:

- markets, shopping centres and retail outlets;
- parks, squares, plazas and other public spaces;
- hotels and unconventional types of accommodation;
- sites related to producing and consuming food and drink;
- gardens;
- science-related attractions;

- features centred on health, holism and the environment;
- locations with a spiritual or religious dimension;
- items related to peace or war;
- outdoor elements;
- activity with walking or bicycling pre-eminent;
- prehistoric sites;
- simple and elemental features and activities;
- attractions focused on modernism;
- cultural aspects (in the widest definition);
- rural and backland areas;
- people encounters;
- items appealing to the senses;
- impromptu happenings and temporary, transitory and provisional features and events;
- places that have fresh appeal because they have never before been tourist destinations.

The most notable aspect about this listing is that this array of features is so all encompassing. Another identifying aspect is the degree to which these items relate to the life of everyday and to lifestyle. This characteristic is the major move from, and contrast with, latter-day tourist products. Many of the items listed would once have been ignored or not regarded as possible to be viewed as tourist products. Some may have been seen as part of tourism but only as secondary or contributory factors, lying in the shadow of a primary attraction. The critical alteration represented by our list is that many items are brought in anew or hitherto only hovered on the edge of tourism but can now be seen as objects that a visitor can be motivated to travel to *for their own sakes.*

Today's tourism is shifting towards more participation, rather than merely observation. Spectators are still numerous, and sightseeing continues as an important activity, but today's onlooker is less detached, more likely to be deeply interested and engaged. Another vital move, and a characteristic demonstrated by many features that are now to be regarded as tourism products, is towards the informal. This aspect shows the quality of dynamism. The element of informality is especially likely in small-scale, folk and community presentations. So, the possibility of participation as tourism providers is offered to those very people often excluded from tourism by their lack of major projects or resources. Since dynamic tourists do not give first priority to clement weather as the central feature or backdrop for their holidays, and since they are also more likely to try a variety of types and levels of facilities (lodging and transport, for example) more places and vast fresh geographical areas are now emerging as credible tourism products.

In the city

Happily, since the world is increasingly urban, cities offer many of the new categories of attractions, and more and more people are motivated to visit these population centres. However, many other elements from this list of novelties respond to our frequent yearning for a complete contrast to urban environments. Many of the products, which represent everyday activities and destinations, are the very components of our routine existence that serve to be the most pleasant, the 'time out', mini-leisure aspects of our daily life. This category of familiar yet special features acts as a bridge, straddling city and rural dimensions.

Cities offer a huge range of products that could be adopted for tourism, and many of these are fluid, temporary attractions. This dynamic variety is due to the myriad of functions and the kaleidoscopic range of people that a city can hold. The city, in its size and different dimensions, offers new, anonymous experiences and encounters to the traveller, fresh formulations, and both conviviality and solitude. In 1974, Jonathan Raban (1974: 15) looked beyond his time and commented that 'The city, our great modern form, is soft, amenable to a dazzling and libidinous variety of lives, dreams and interpretations'. He spoke also of the 'special relationship between the self and the city: its unique plasticity, its privacy and freedom' (ibid. p. 250).

We have already explored *flânerie* in Chapter 5. We have also considered how this urban pedestrian activity of strolling can be a prime feature of tourism. Among objects of interest and places to venture into are department stores and shops. Donald Horne (1992: 176) describes how people who view galleries, centres and display salons of Seibu in Ikebukuro in Tokyo develop 'museum eyes' and how, once the visitor sees from this optic, 'it is then easy to turn the rest of the store into exhibits'. A shop has an innate, dynamic quality; it has a continual influx of new goods, and its displays change constantly. When there are a number of shops together in one street or shopping area, the group dynamic escalates the amount of frequent change to be viewed. Moreover, the actors in these areas are altering constantly and so a constant renewal of diversion and interest is provided. Looking at things from this perspective, it is difficult to see how a museum, usually in an overall greater straitened circumstance, can compete on this basis with a vibrant shop or shopping district.

Out of doors

Markets offer much the same possibilities as a group of shops but, because their whole flavour and atmosphere is to be temporary, fluid and impromptu, they are even more dynamic in character. They are, of course, also often *out of doors*, a characteristic that renders them as accessible as the streets, and frees them to showcase a greater mix of players. Their original role naturally gives markets structure, another asset as tourism destinations. Their framework is sufficient to allow them to be managed to achieve set objectives, and control can be exercised over them as required. In certain aspects, parks and plazas can also offer this same useful dimension, since there is a measure of control in their make-up, but the degree of control varies with specific sites. Plazas formed by office buildings or in business corporation ownership may allow public access only under tolerance, to maintain good public relations, and they may also be available in this way only during normal working hours. Squares range on a continuum from the extremes of completely private use to being public and entirely open. The overall context of these places is to be scenery, permanent in nature, and to deliver the option of being a backdrop for a range of other presentations that may be temporary. While playing this role, these urban spaces can serve as important locations for random encounters.

Accommodation, and food, drink and hospitality

A new attitude among tourists is to view accommodation as a destination and reason to travel. The move was encouraged by the emerging concept of 'designer' hotels. Across the river from Bordeaux, the Hotel Saint-James, built according to the formulation of the architect Jean Nouvel, took the lead in this movement. Other innovative destination hotels include the Wasserturm in Cologne, which is a conversion by Andrée Putman of a water tower. In New York hotels such as the Royalton have been provided by entrepreneur Ian Shrager with Philippe Starck as his favourite designer but with the initial example, Morgans, built in 1985 to Andrée Putman's design. The first essential target market for these establishments was the fashion and design leaders and groupies, as well as celebrities and sophisticates. It was a natural extension of Schrager's empire that he should cater to his chosen market by opening the Delano hotel alongside other fashion hotels in the revamped and newly-trendy Miami South Beach, and the Mondrian in Los Angeles.

London is now home to a small collection of Schrager hotels, beginning with the St Martin's Lane opening and continuing with the Sanderson – a refurbished showroom for fabric, wallpaper and paint. These hotels meet competition in this capital city from the Hempel by designer Anouska Hempel, the Halkin, the Metropolitan, and the Great Eastern, a grand old railway hotel at Liverpool Street Station through which Sir Terence Conran has entered the field to try out the general genre. Nouvel has added a new wing to a restored historic building and has thus delivered to Avignon in Provence a designer hotel, the Cloître Saint Louis. Clones of the designer hotel type are now to be found around the world – from the Clarence in Dublin to the Regents Court in Sydney. Some of the most luxurious of these are those of the Aman Group, whose residences stretch from Jackson Hole

*Hotel as destination: the Jean Nouvel wing
at the Cloître Saint Louis, Avignon, France*

in Wyoming to Bali. A hotel of most intrinsic dynamic nature is the cele-
brated Ice Hotel at Jukkasjärvi in northernmost Sweden, portions of which
are created by carving them afresh from ice each winter season.

A development to the designer hotel phenomenon in Great Britain is
where this select type of accommodation has metamorphosed to become
approachable, middle and main market, and established in areas outside
London. This development has come about essentially in the Malmaison
hotel concept, which began in Leith, the port of Edinburgh. From Leith, the
concept continued to Glasgow, and eventually a series of several hotels was
opened and then sold to a large organisation to became almost ubiquitous
in post-industrial areas, such as the cities of Birmingham, Leeds, Newcastle
and Manchester. The Malmaison chain is now destined to come to London,
and to spread further afield, to the European mainland. Malmaisons make
a feature of their restaurants, so there is a link for travellers with the 'food
outlets as destination' role. Conran has made a major contribution to this
formula by designing a large group of eating establishments, some of them
'restyles' of buildings or rooms formerly devoted to other uses.

The Estate Concordia 'ecotents' at Saint John's Maho Bay on the US
Virgin Island of Saint John may seem to be at the other end of the accommodation scale from luxury hotels, but nevertheless they are habitations
designed for the traveller of discernment. These 'ecotents', each housing six
persons, are set on stilts to avoid interrupting the foliage, and they depend
on solar technology (Siebert, 1995: 89; Pierce, 1995: 89). In the French rural
village of Mens, enterprising residents realised that their area would be an

attractive location for walking holidays, but recognised the difficulty of providing adequate accommodation in the village, since the old local hotel was outdated. In order to preserve the rural lifestyle and keep their village in existence, while developing an economically viable tourist attraction, the villagers adopted a strong environmental focus, choosing to become an *écopôle*. To their own funds villagers were able to add outside assistance, both from within France and from the European Union, to build a new hotel. Alongside the village is a 'Living Earth Centre', directed towards 'green gardeners'. The project's environmentalist instigators are planning for 'a day when thousands of visitors will study their bizarre vegetables, wood burning furnaces and mud huts' (Sage, 1995: 16).

The attraction of huts and other uncomplicated forms of cover is a typical aspect of Dynamic Tourism, part of the inclination to simplicity, flexibility and less permanent structures. As with the case of Stockholm, when playing the role of a European Cultural Capital (see Chapter 4), Glasgow delivered a hut exhibition as an event during its West End Festival in 1999, when the city was hosting the United Kingdom Year of Architecture and Design. This 'Ideal Hut Show' represented ideas drawn from celebrity architects and designers for adding on to a basic garden hut.

In London, in the same year, the 'Paper Log House' was displayed, as part of the unusual and dynamic 'Cities on the Move' exhibition (designed by Rem Koolhaas) at the Hayward Gallery. It was placed on a Gallery terrace, in juxtaposition with nearby concrete and tower block architecture and a neighbouring 'Cardboard City' of the homeless. The 'Paper Log House' was constructed of cardboard tubes and beer crates, according to a design by Shigeru Ban. Its dynamism is revealed in that not only was a house in the exhibition, but earlier examples had been employed as temporary housing for homeless victims of the Kobe earthquake. It is reasonable to suppose that dynamic tourists, in their wide perspective will have an interest not merely in designer models of the type, or in viewing the idea, but will want to see the world's home-produced equivalents in their barrio and favella social environments.

We have already noted that tourism directed toward food and drink is gaining prominence. With the exception of the necessary meals consumed while on holiday, and the existence, as mentioned in Chapter 4, of such forerunners as 'whisky trails', food and drink aspects have been only minor and attendant attractions. Along with generally greater tourist motivation toward gustatory features, the focus has been on public concerns and interests with 'real', 'back to nature', and organic food and drink. Dynamic tourists, with their sophistication and broad interests, are likely to be amongst the strongest enthusiasts. Markets serve part of the particular interest in food and drink, but a tourist's interest is often focused on seeing the home of each item. The places of origin of comestibles or drink are popular with visitors; an example might be the place where a particular type of farm-house cheese is produced.

Wine has a particularly prominent spot in the new focus. The places of origin of the finest wines, such as the châteaux of Bordeaux and Burgundy, California's Napa Valley and New Zealand's Marlborough vineyards, are appealing places, though in some cases, as mentioned in Chapter 2, their local landscapes are indeed rather dull in themselves. As also mentioned in Chapter 2, the importance of wine landscapes in cultural terms is confirmed by the recognition at the end of the millennium of the wine area centred on the Bordeaux town of Saint-Émilion as a World Heritage Site. Self-guided tours of wine regions give tourists the responsibility of choosing their own route as well as a heightened responsibility to express their own individuality and knowledge in seeking the potential pleasure and satisfaction of selecting and buying quality wine.

A number of indications bears witness to producer recognition of this burgeoning public interest in wine, and the desire for education about it. The visitor attraction of 'Vinopolis' in London is one example, with retail outlets and restaurants as part of its large complex. The *Touring in Wine Country* series of books testifies to our interest in travelling specifically *for* wine. The editor, Hugh Johnson (1997: 7), explains our fascination with wine, 'Wine, more than anything else that we eat and drink, is resonant with the sense of place. ... No wonder wine tourism is growing.' Back in 1967, Marshall McLuhan described wine as 'supreme in oral societies' and reported that wine's first appeals are sensory and direct. He warned against, 'wines ... being talked about but never consumed' (McLuhan, 1967: Item 18). From the amount of wine currently being imbibed, it appears that McLuhan's fear is unwarranted. This oenophilic tendency fits dynamic tourists' formulation of the need to address the instinctive and immediate dimensions of the senses and also to honour intellectual dimensions of experience.

The garden movement

We have already emphasised the place of honour held by gardens in Dynamic Tourism; the attraction of gardens for tourists is an instance of sensory appeal, since gardens engage our faculties of sight, smell, touch and hearing. Gardens are like wine in this way, and also in the interest generated by information about their creation, ownership and care. We are tempted to view any gardens of excellence and interest, such as those illuminated by the United Kingdom's National Gardens Scheme, as well as those star quality gardens with international reputations. Garden centres must be counted among those items that ought to be recognised as visitor attractions, and that have great potential for growth in that role, as long as they can show persistent high quality.

Moves in time

Since sensory and experiential factors hold such important positions in Dynamic Tourism, a large number of new tourism products characterised by these properties are now being defined and increasingly recognised as prominent. In Chapter 6, the contemporary trend to visit popular destinations outside their usual hours was defined as a dynamic difference in tourism patterns. We considered this question specifically in connection with the availability of tours of Antwerp in the early morning. In fact, the deepest and most fundamental character of a place may be revealed more clearly in these off-peak hours. Consider the possibilities in visiting the commercial district of a city early on a Sunday morning, when its busy-ness is removed. Visitors to the City of London at this particular time, when a less-peopled quietude reigns, may observe an essential skeleton of structure and may more easily notice details and defining elements not easily sensed during the workaday hubbub. Then, if ever, the sensitive visitor may hear the ancient 'voices' of an old city.

Spiritual moves

The recognition and search for an accurate and full sensory experience of travel destinations may also link us to products whose essence is their spiritual dimension. Such products, as we have stated, are also important features in Dynamic Tourism. Among the many sites and activities linked to formal religious practice, the items that are generating special appeal today are those characterised by simplicity, by exoticism (that is, items related to religions other than the visitor's own) or, paradoxically, are those of which the visitor has the best knowledge and familiarity. What gives these sites their appeal is their novel elements, together with their intrinsic spiritual qualities. For this reason, we should include places outside the traditional definition of formally religious sites. Indeed, such places may perhaps form the major sector of this category for dynamic tourists.

We have already discussed the importance of the spirit of place as an attraction for dynamic tourists. Among formal Christian religious institutions, Cistercian monuments, such as those in the South of France at Sylvanes, Silvercane and Thoronet, are especially suited to predilections for simplicity and spirituality. As a whole, monastic sites tend to demonstrate these characteristics, even though their exterior architecture is sometimes heavily ornamented. A representative case is the Spanish Franciscan Mission Church, 'Queen of the Missions', at Santa Barbara, California. The church began as a simple adobe construction, and grew to monumental size, decorated with eye-catching ornamentation. However, the typical missionary's bedroom on display for visitors inside the Mission complex is extreme in its simplicity. The interest of the church is increased by the monastery cemetery, which joins the graves of the missionary Brothers with the community of Chumash Indian Christian converts.

Moves to death and commemoration plots

Cemeteries and graveyards, strange as it may seem, are 'growth' items for Dynamic Tourism. They draw interest by the people commemorated within them, of course, but they are also intrinsic sites of contemplation. When cemeteries commemorate war dead, they serve another further purpose, in reminding humanity of the horror of war and the necessity of peace. Since many major battles of the First and Second World Wars took place in northern France, there are many cemeteries for the fallen there. Some are small or are in locations that are rarely visited by tourists, such as a French national cemetery near Remy, where the honoured dead of continental France, their graves marked by crosses, are joined by men whose monuments indicate origins in an Islamic homeground.

Modern moves

In recognising the attraction of simplicity, elemental features and the mystery lent by great antiquity, it is logical to predict a move toward greater tourist appeal for prehistoric sites. Maybe the general aura of millennial mystique added an extra lustre to the growing power of this appeal? However, a paradoxical corollary of the attraction of prehistory and the primitive is a growing interest in the Modern Movement, founded on emulation of the primitive in its simplicity and elemental strength. Edifices of the Modern Movement are now emerging as sites with increasing visitor interest. These sites can now be perceived as heritage, since they are from the last century and are some generations old. This interpretation probably makes them more appealing in some quarters than they were earlier. Before the century's end, the National Trust had acquired several examples of the Modern Movement in England. High Cross House in Devon, distinguished by its International Style, was restored and opened to the public by the Darlington Hall Trust. The Trust's archivist (Gillilan, 1995: 60) is quoted as remarking, 'People still dismiss the house as ugly ... [but] almost everyone alters their perception once they step over the threshold. It's like being in a Mondrian painting'. The Villa Savoye, one among a group of private residences in and near Paris that were designed by Le Corbusier, has been restored and opened to the public as a participant in the French patrimony because of its location, although the architect was Swiss by birth. Mies van der Rohe's reconstructed Barcelona Pavilion, ultra simple in style, seems poised for more wide appreciation; it is conveniently located for tourist appreciation among the suitable environment of the city of Barcelona. We might note that the overall increase of interest in things modern, with their deceptively simple materials and minimalist structure,

Taking in the Modern: Mies van der Rohe's
Barcelona Pavilion, Barcelona, Spain

as well as their emphasis on the ephemeral, adds to the move by providers to present more temporary forms of simple shelter, such as tents and cabins, as tourist attractions.

Moves to the natural, isolated or untamed

As more people lead increasingly urban, communal lives, an increasing number of new tourism products cater to their need for a contrast to their everyday existence of constraint and superficiality. Contrast is a strong element in holidays that focus on spiritual enrichment, self-actualisation and individual refreshment with a sense of liberation. Many of the products already presented here as forming a part of Dynamic Tourism respond to this demand for 'oppositeness'. They demonstrate features of spirituality, appeal to the various senses, and inspire contemplation. Rural, unsequestered places and those items that deliver a feeling of going 'back to nature' and fundamental experiences are among the products eminently equipped to meet this defined need.

In this connection, backcountry and other under-visited areas are potential tourist products. Part of the Périgord in France, the 'unexplored heartlands of the Dordogne' show a recognition of this human impulse and the possibility of attracting more visitors, in a certain and special way. Here, 'the authorities have woken up to a new form of "green tourism"' and it has been deduced that 'what visitors really want is an initiation to the joys of simple country life' (Street, 1995: 20). In England, the county of Northumberland represents the country's most remote and uninhabited portion, although it is far from being a genuine wilderness. However, the Northumberland National Park sees its relative isolation as an attraction and uses it to promote itself as 'the land of the far horizon'. It shows a sophistication and empathy for its target market in remarking, that 'wilderness is a spiritual concept rather than a reality' (Northumberland National Park leaflet, 1992).

Moved to walk

One implication of the definition of a tourist product along these lines is an emphasis on walking. This activity has already been shown to be a paramount feature of Dynamic Tourism. The notion of freedom in walking is used by the Ramblers' Association, newly calling itself just 'The Ramblers', both in a main slogan 'Feel Free' and in offering the corresponding opinion fhat 'Going for a walk gives you a unique sense of freedom' (1995 leaflet). The Ramblers' style of promotion assumes that walking for freedom occurs only in the countryside, but of course, as depicted in Dynamic Tourism, a similar emancipation is found in strolling anonymously along city streets.

Moved into events, and by fluidity and temporary occasions

If traditional tourism emphasises a set menu of monumental sites and itineraries, Dynamic Tourism is characterised by its fluidity, and this is demonstrated in the recognition paid to satisfactions found in the impromptu, temporary and ephemeral. Temporary items, such as events, festivals, 'happenings' and occasions, may seem spontaneous, but they must be exactly and minutely planned by their providers. One popular ephemeral event with a firm structure, for example, is the television show with a live audience; the public is eager to attend and their presence enhances the show's effect. Even for unplanned events, the context in which they are likely to occur can sometimes be premeditated and delivered in such a way as to also deliver the 'uncontrollables'. Indeed, it is often necessary that it should be so, both from the provider's point of view and for the viability of the event. For the visitor, that a feature is unexpected can give it greater value and memorability. For example, the unorchestrated glimpse of a swimming beaver may be the single most memorable image from a whole visit to a rural area of Canada.

Among the category of fleeting visitor attractions are music concerts and performances. (This is except, of course, that they may be recorded and so may be repeated often, and to a wider audience.) The most fluid, flexible and versatile type of these temporary products is a jazz performance. After all, the essence of jazz is dynamic improvisation. Tourists visit New Orleans, go to jazz festivals and attend Michael's Pub in New York on Mondays to hear Woody Allen on clarinet. In spite of all these activities, however, jazz has yet to 'take off' as a major product of tourism.

The new style of communications technology offers predictability with variability, customisation and mobility. The 'virtual Xanadu house' belonging to Bill Gates, the CEO of Microsoft, demonstrates the potential for communications applications. A ten-minute canoe ride across Lake

Washington from Seattle, the Gates home boasts digital walls. It is reported that 'Instead of travelling the world to collect great art ... Gates has purchased the electronic rights to art from museums. ... With the press of a button, the bathroom walls will become works by Rembrandt' (Egan, 1995: 24). In this example, replicas are not only made to serve the same purposes as originals (but with infinitely more flexibility), they can also be efficiently gathered up into collections and displayed at the locations most conveniently accessible for the viewer, including private homes.

The New Style on Display: The Earth Centre

The Earth Centre, mentioned in Chapters 3 and 6, is a new travel destination that embodies many aspects of Dynamic Tourism, and that serves as an indication of the style to be sought in tomorrow's attractions. Its mission is education, its subject, the all-embracing study of the Earth, its aim, to encourage new perspectives for future activity; 'You'll never see the world in the same way again'. The Earth Centre represents re-use in two major ways: its location is the site of two former mines and it is a 'new kind of theme park' and thus is adapting a concept that exists and using it in a different way and towards its own particular objectives. Everything at the Earth Centre tries to keep within the theme of showing action towards sustainability, of conserving resources, of demonstrating sensitivity and of being generally caring towards the Earth and its people. The Earth Centre looks for a 'green economy'. It is an initiative funded from Millennium Commission lottery funds, the European Union and the development agency English Partnerships.

Visitors to the Earth Centre café may enjoy organic cuisine, accompanied by the background music of recorded bird songs. The core of the Centre is the Action for the Future Gallery, but there are also the sound and light movement, Planet Earth Experience, and a shop and advice centre. All these elements are indoors. Many of the attractions, however, are outside, viewed as the visitor strolls through a variety of landscapes, gardens of experience and sites of explanation. Communication is achieved through sensory stimulation, experiential and interactive displays. Exhibits very often tend toward the abstract, but there is a very concrete example of the principles embodied there in the fact that the Centre's waste is visibly treated throughout the site.

Yet another potential attraction for dynamic tourists is the Rokkaku Trail that 'brings all your senses to life'. Here, the sky can be heard through large trumpets. Visitors walk unshod over changing types of ground material. The yurt, a traditional form of round tent used by Turkan and Mongolian nomadic peoples, is on display. So is the Kaki tree, raised from a cutting taken from one of the few trees that survived the nuclear bombing of

Nagasaki, at the end of World War II, and in itself a very potent statement of Dynamic Tourism's preoccupation with peace. Across the wider Ecology Park is the Trans Pennine Trail, which is linked to the National Cycle Network. Admission costs for the Earth Centre are reduced for visitors who arrive by bicycle, boat, bus, train or on foot. Staff members, be they actors, interpreters, rangers, gardeners or the people who man the shop, advice centre and restaurant, are expected to be open to approach from visitors for impromptu, dynamic dialogue, ready to dispense information and answer questions. The Centre stages special events, with an overall theme of encouraging visitors to develop their individual responsibility and knowledge for the good of humanity. Bad weather need not be an impediment to visitors going outdoors during a tour of the Earth Centre, because natural factors such as rain can be regarded as mere extensions of the presentation and the visitor's sensory experience. The Centre also publishes its own magazine, *unearthed,* for visitors who wish to maintain a connection after their stay is over.

In some ways the greatest challenge to the success of the Earth Centre is its location, as we discussed in Chapter 6. It still remains to be seen whether the Centre will be able to demonstrate that its novel and unstereotypical location, which is an integral part of its process of innovation and change, is no longer an impediment to attracting a sufficient number of visitors, in spite of traditional provider perceptions about what is necessary to sustain tourist volume. Interestingly, Phase Two of implementation includes standard mechanisms to help bring increased use and attention to the location; these being a conference centre on site and an associated hotel.

Moved More by Experience

The tourism products mentioned in recent paragraphs demonstrate the difference between contemporary attractions and those portrayed early in this chapter as traditional. They show that a fundamental change to new products is occurring. One defining aspect of many of these new products is that they are *experiential*. This allies with departures in the activity of shopping, itself a tourist item in many instances. New departures in shopping as a tourist activity are identified by B. Joseph Pine and James Gilmore in their book, *The Experience Economy: Work is Theatre and Every Business a Stage* (1999), and further described by Lawrence (1999: 10) who cites Southwark's Vinopolis as a prime example of the selling of experience.

Individuals and Their Senses

In new products, the focus is much upon individuals. There is an effort to create connections with their senses, enhancing their being and personal development, while remaining in connection with the world and its issues and elements. One example of how the tourism industry is awakening to some of the trends we have identified is this promotion by Kerala Tourism for its hill resorts:

> Walk into our spice plantations. Trek up cloudy hills. Listen to the trumpet of elephants and the roar of tigers. Hear flying squirrels whizz by rustling leaves. And birds sing along gurgling waters. Droning beetles, chirping crickets, the soul of life. Just two hours away back-waters whisper, seawaters thunder and ayuvedic regimens encompass the senses. You will want to keep on hearing our bands. (Kerala Tourism advertisement, 1999)

The West Virginian resort, conference centre and spa, Coolfont, presents a related example 'in a wilderness atmosphere'. It positions itself clearly, recognising contemporary strong individual preferences for personal choice and responsibility in the affluent target market. It recognises and validates the urge to escape to the wild, to explore sensual stimuli and to experiment with New Age and alternative therapies. To the busy, stressed potential visitor, the 'pitch' is:

> You deserve some time for yourself. Coolfont's beautiful 1350 acres of woods, trails and lakes is the perfect environment to relax, rest and renew. Enjoy soothing health services such as massages and facials. Rejuvenate with aerobics, classes and hikes. And connect to a quiet peaceful place within as you experience yoga, meditation and tai chi. Choose structured programs or do nothing at all. Choose well being, choose relaxation, choose to take care of yourself – choose Coolfont. (Coolfont advertisement, 1999)

Environment, Peace, Tranquillity and Sensuality

Environmental topics and peace stand out among the global concerns of the new tourism products in general. The connection is through the individual, who is invited to become informed and then act accordingly. Peace features as a needed individual aspect, as well as the object of pacifist desire for the world in general; in its individual sense, it appears in the combined phrase 'peace and quiet', frequently in connection with the spiritual and/or religious dimension. Travel serves to bring individuals to new encounters, to sensory, spiritual connections of calm or stimulation, of connection or solitude. Extensions can be made, and must be made, to fresh places around the world. The sheer surprise of difference for its own sake delivers interest to the beholder. There is a general widening of view. The simple, direct and elemental forces of the world around us are often brought to the forefront, both as products offered to tourists, and in the way travellers consume those same products.

Shifting, Passing and Widening

Of the fresh offerings of contemporary tourism, many are transitory in type. Frequently these offerings are 'light', not monumental and, often, not even material. With many contemporary product types, their existence itself is not new, but their novelty lies in being perceived, finally, as potential attractions for tourists. In those cases where the offerings have already existed as tourist products, the move, and the newness, are in refashioning or repositioning them in reaction to fresh tastes in the target audience.

The overall essential move that this chapter has sought to define, is that of widening the range of items in the tourism domain.

Chapter 8

Redeploying

Tourism Products for Renewal and Change

Dynamic Tourism fills the need for a wide and flexible choice of tourism products. The more dynamic new products there are, the less will be the tendency for tourism to produce crowds and 'peak season' inconveniences at the same old list of traditional destinations. When does the tourism industry need to add more attractions? Action is needed where there are too many tourists at a given site for them to experience it with pleasure, where a site is over-run with visitors to the point that its physical fabric is over-stressed or its support structure and host community are put under too much strain, and also when other potential sites wish strongly to entice tourists to visit their locations.

Some tourist destinations that have enjoyed popularity in the past, even though they still wish to remain in the tourism market along with the new, larger portfolio of products, may suffer from a reduction in their appeal, or may find themselves rejected by today's tourists. Either in reality or in the perception of contemporary travellers, these sites have 'had their day' and are seen as dull, stale, and outdated.

Other old tourism sites, now partially abandoned, are being re-imagined, re-defined to shift to some new tourism aspect, changing from their past roles to new ones, with a fresh dimension placed in the limelight.

Yet a third group of old sites are the 'white elephants' of the tourism market that need redesign and development because, even from their first formulation, they were not profitable or appealing.

A different category is sites that were not designed to appeal to tourists at first, but now face a future of needing a new purpose, and tourism could serve as this. Maybe the original economic base of the community is shrinking and a new role is needed to keep the area vital. Or perhaps the changing tastes of contemporary tourism offer a chance to redefine and appreciate those facets of the area and community that tourists of the past might not have found enticing.

All these are examples that call for change and redeployment. Old sites that were once defined as tourism attractions and developed to appeal to a certain market, now need a new definition and development, or else an old site once devoted to another purpose must be altered for a new existence as a tourism product.

Perennial in appeal, but in need of attention

Long-standing products do not necessarily lose their appeal as time passes: tourists still flock to see the monuments of ancient Greece, Egypt and Rome or to visit Notre Dame Cathedral in Paris, as they have done for centuries. The power of attraction of traditional tourist products is often so strong as to override a weak provision for visitors in terms of management, presentation and interpretation. However, many such older, traditional form, features and activities are now either actually losing visitors or have reached a plateau of future potential downfall and are operating under circumstances that mean that the contemporary requirements of their tourist and provider are not being met.

Stagnation or decline: out of touch or connection

Products that have been offered to tourists throughout a long period of time, may fall into a state of stagnation, or they may decline owing to one, some, or all of the following reasons:

- the complacency of providers who do not continue to adequately promote their existing attractions or who never present their offerings in a new light;
- the failure of providers to keep delivering new objects of attraction;
- lack of tourist interest in the features on offer, or failure to see current relevance and appeal in them.

Generally speaking, the underlying reason for such a reduction in visits is that products do not adequately respond to the concerns and wishes of their tourist audience, and thus of society overall. To this we may add another possible cause, the cruel blow of fate when an attraction loses its connection to the larger world, through the closure of a connecting railway, loss of a bus link, or cessation of flights to the nearby airport. Often these closures and dis-connections of public transport links are beyond the immediate control of the place and its community. Being in a situation where a once-suitable site has changed and is no longer appealing, owing to changes in trends and habits in society, is a difficult matter to overcome. Tourism practices have changed significantly, travellers visit different places and types of venue than they did in the past, and many nations that once did not send tourists off for foreign holidays now send an increasing number of voyagers to increasingly more distant destinations.

Revivals

It should be emphasised that old products may have a current life in exactly their old role. After suffering a downturn and falling out of tune with the tourism market, they can re-emerge as suitable with their original features. However, it may sometimes be necessary to demonstrate explicitly that traditional items have gained new relevance; contemporary audiences may not notice these aspects without the assistance of new explanation and promotion.

Lifecycles and Interventions

It is precisely those changes in tourist tendencies and preferences, recognised by Dynamic Tourism, that make possible the transformation or reaffirmation of all these kinds of entities as viable tourism offerings. The essential steps in the process are:

- recycling or re-energising items;
- deploying features which have not been identified as for tourism before;
- finding new applications in tourism for products that have become under-visited.

In essence, Dynamic Tourism expects a future that calls for constant re-evaluation with attention paid to keeping existing items in the arena as tourism products.

Expecting a product to continually perform on an upward course, or even at the same level, is optimistic. Part of the fluidity of tourism, which is recognised by Dynamic Tourism and led to its emergence as a concept, is due to our contemporary tendency to seek ever-new travel ideas and sensations. Existing products must either change continually to keep the attention of a tourist market, or else must be promoted to fresh new markets; the alternative is to accept the inevitability of waning interest among tourists. The present chapter seeks to suggest ways to avoid downturns, while also demonstrating that a downturn need not be the end, and can be followed by the creation of an upward turn.

Plog (1994: 48) has described the impact of what happens when a destination changes and loses its original characteristics of attraction. While tourists scarcely regret the change and slump in tourism, because they merely move on to some new place that possesses the characteristics of appeal, the effect of this alteration upon the general locale and the local residents is strong. Plog says:

> As the tide turns against a destination, from glory days to dreary days, the process can be relatively quick. Travel fits in an economic climate that is much more competitive than was previously the case, and the new socio-economic realities can contribute to a very rapid fall from grace of the travel spot. (ibid.)

Keeping in Close Contact with Tourists' Tastes

Pessimists predict the inevitability that every product's appeal will wane. This book maintains that such a downturn is a possibility, not a certainty. Whether we meet with decline and rejection rests heavily on the ability of providers to serve and present their product in a responsive and flexible way, in order to maintain visitor attention and interest. Travel writer Frank Barrett provides an example of how tourism providers lack understanding of a change in holiday customers, in contrast with the escalation of product by British supermarkets in response to changing tastes among consumers:

> To a large extent, this general up-market shift [by supermarkets] has somehow evaded the major holiday companies. The sun-and-sand brochures of summer 1998 are largely indistinguishable from the sun-and-sand brochures of 1968.
>
> The charter flights are now in bigger jets, but the hotels are still largely the same anonymous tower block places which continue to offer the same unimpressive accommodation and the same bland Euro-food. Of course, there is a much more exotic range of holiday destinations on offer these days: the Dominican Republic, for example, or the Gambia. But in reality, these places offer the same bland package, only in a more tropical setting. People deserve better. People want better. (Barrett, 1998: 76)

Subjects for Redeployment, New Encapsulation, or New Inclusion

In summary, those items open to redeployment are:

- Old tourism products that are suffering from neglect or that have lost much of their appeal.
- Old tourism products that need a different aspect to be developed for tourism.
- Products originally aimed at attracting tourists, but which have never succeeded in this aim.
- Existing entities that have potential for a fresh use as tourism products.

Of these four groups calling for attention, most are old tourism products that need to be changed, improved or reoriented to a different, appealing dimension in order to survive. One member of the group, however, contrasts with the rest, since it is made up of old items that may survive through becoming tourism products. To a certain extent, this last group is the easiest to deliver in a form suitable for contemporary tourism. Apart from their intrinsic elements, many or all of which must be retained because of heritage preservation strictures and other criteria for retention, the group's participants are *tabulae rasae.* They are new as tourism products. The other categories are handicapped by the fact that they all now have some shortcomings as tourism products. They are currently deficient in appeal, even though many have had a measure of success in a tourism role in the past. In all the categories defined here, however, there is one feature that all the sites share; they are all old, whether in concept, role or physical plant. One way or another, redeployment is needed if these destinations are to serve in today's tourism.

A redeployed site, and a day-one success: Tate Modern, the public, and the Tate Gallery Director, Sir Nicholas Serota, await first opening time, Bankside, London, England

Dynamic Tourism is based on tourists who are intelligent, able, and eager to see a full range of new and different products. These tourists are presumed to have an open mind about reviewing and revisiting old items, as long as these are presented to them and shown to be of contemporary relevance, interest and appeal.

The kinds of entities with need or potential for redeployment in the tourism context range from parks and gardens to seaside resorts, spas, shopping galleries, malls and arcades, museums, means of transport, and industrial monuments. We should note that these attractions often date from the nineteenth and early twentieth centuries. While in most cases their physical fabric stands in need of refurbishment, the major point in terms of tourism potential is that these sites need functional and/or conceptual redeployment. Of course, it should be said that any time period could furnish us with old sites whose original function was not related to tourism and which have the potential for retooling to attract visitors. The industrial period is mentioned specifically above, because it tends to deliver especially large numbers in the category of changed use, and also because the necessary alteration for the item is a radical shift from work entity to leisure.

In the United Kingdom, the Liverpool Tate Gallery, housed in a portion of the Albert Dock (a Grade One Listed Building) is a particularly good example of an old building re-fashioned for tourism. The Tate Modern on Bankside in London and the BALTIC on Tyneside are other major modern art centres that illustrate this type of reworking of an industrial site to new uses. They are massive examples, since each has undergone radical alteration in its features to fit its new function and become a visitor destination. Slightly different examples are presented by shopping arcades. For instance, the Royal Victoria in Sydney has not altered its essential role as a shopping venue, but it has been changed and embellished in appearance, turning an attractive new face to the world, in order to suit tourists as well as local shoppers.

Changed and restored shopping and strolling: the Royal Victoria Arcade, Sydney, Australia

Out of taste, place or fashion

Let us now consider entities that already have a history in tourism. If a decline in visitors has marginalised a formerly successful product, an evaluation of the situation is needed. Why has it fallen out of favour, and what can be done to restore the item to enable it to survive as an element in the tourism marketplace? A careful study is a necessary preliminary to determine the best type of attention to devote to putting things back on the right track.

As we have said, some destinations are caught in the wrong geographical position: changes in public transport leave them inaccessible or patterns of tourist behaviour alter. There may be nothing intrinsically wrong in the products themselves; they are simply not in the best location for the present-day market. Attention to 'turning them around', therefore, needs to depend on addressing and altering the surrounding circumstances, or on changing the perceptions of potential tourists about those circumstances.

Other destinations, such as certain 'grande dame' resort hotels (for example, those in the Adirondacks of up-State New York), inland spas and holiday camps may be out of step with contemporary public taste as entities. They are also tucked off the beaten track as locations, again because of alterations in tourists' travel choices or the reduction or removal of public transport facilities. In general, old, under-visited resorts now need to answer one big question: are outside circumstances the main reason tourists have lost interest in them, are there individual causes within the particular sites, or does the problem lie in a combination of both kinds of factors?

Quality, Standards; and Presentation

If a product has once enjoyed success, there must be some reason to explain why visitors have relinquished it. It may be tourist boredom, coupled with the contemporary expectation of novelty or some other factor. One likely cause is a provider failure to deliver the increased levels of quality required and expected by contemporary visitors. Meanwhile, competitors may be providing what is needed to satisfy today's increased standards.

Another major cause, perhaps *the* major cause, for rejection of products by the tourist market, is provider failure in *presentation*. This lack of sufficiency may be in inadequate maintenance of the physical substance of an item. On the other hand, provider complacency may interfere with presentation of a product to the public on an on-going basis, and the provider may neglect to showcase any new features. Museums, as a group, are sometimes prone to this latter failing, in that, through arrogance or simple lack of understanding, they do not bother with the task (Boniface, 1998a: 25–32). When, due to a reluctance or an inability to invest the funds needed, too little care is taken in presentation to the point that the treatment of the physical fabric of the tourist destination suffers and facilities look worn and take on a tawdry air, the site may be self-condemned to visitor loss. Shabbiness in the facility will repel potential tourists.

As we address the question of presentation, the wider aspect to be defined concerns the choice of whether an existing 'persona' or public identity for a facility will serve well as its contemporary concept, or whether other facets of the item need to be focused upon, or fresh ones created. The matter is difficult, and requires careful thought. The first impulse in a situation where there is a rapidly spiralling downturn can be to try radical alteration. Yet, bearing in mind the inclinations and emerging needs of dynamic tourists, just the opposite sort of action might be needed. Old elements may have exactly what it takes as items of new enticement, albeit that they must be well and recently refurbished and that they will probably also need fresh presentation to a market.

The fact that a tourist attraction is situated in a 'backwater' location, and thus contrasts with routine elements, may actually be its major factor of appeal. Indeed, the features that differentiate an entity from the competition equip it for a role as a holiday product because of this very contrast with the standard features of everyday life. Hotels, for example, are now very much alive to this factor and many 'old glories', now equipped with the accoutrements expected for modern day travellers (modem ports and indoor fitness centres, for example), are experiencing a resurgence of visitors. In their idiosyncrasies, such hotels offer a welcome alternative to the sameness of chain and franchise hotels.

Old museums share this dilemma of whether or not to stay as they are, while perhaps showcasing the historical aspect more, by showing old collection presentation styles as a feature, for example. Their alternative is to revamp themselves in line with the latest contemporary surveys of visitor preferences. After all, museums might well seem to be outmoded institutions, since there are now more modern ways to communicate information about bodies of specialised knowledge, such as the past, the natural world and science. However, the key particularity of museums in their traditional format has been their collections, the core of their existence. MacDonald and Alsford describe the dilemma faced by museums and depict their general situation:

> Museums find themselves in a difficult situation. ... It is no longer sufficient for museums to centre their attentions on artifacts and the static displays in which artifacts are usually presented. Expectations of museum visitors and the tourist population generally are rising, partly because of other cultural or recreational institutions with which museums are in competition. Visitors continue to seek both educational and quasi-religious experiences in museums, but they also want to be entertained, to have their senses stimulated, and to be offered comforts and conveniences not previously found in museums. (MacDonald & Alsford, 1989: 45)

Such are the problems and choices available in updating museums. One way of revamping, which has been adopted by some museums, is to keep collections displayed in the traditional format, but to keep maintenance at the very best current standards, thus demonstrating the mind-set of collecting that prevailed 'once upon a time' when the museum was first established.

New Faces, Roles, Positions and Interpretations

The seaside: a special case

Old seaside towns and resorts are among the items most in need of action to bring them back into popularity. They are also sites with the greatest potential for revivification. They share certain characteristics with spas, although, if the spa is not at the seashore, it lacks a key asset. By definition, the seaside is marginal, 'on the edge'. This fact seems to impart some intrinsic feeling of greater freedom in people, a feeling that is lacking in land-locked resorts. The shore provides an implicit environment for personal expression.

Bath, the old British spa town, is an exception. In its heyday, its role was to offer its clientele, 'society', the rich and their hangers-on, as much freedom from the strictures of home as any seaboard resort. The city is now a World Heritage Site, in recognition of the stylistic uniformity of architecture in its eighteenth and nineteenth century buildings. Generally speaking, Bath has revived itself for visitors and locals through the re-delivery of its old styles of attraction, such as shopping, concerts and theatres, as well as the spa itself, but these old attractions have been reformulated to meet contemporary inclination.

Beaches have a long and varied history of human use (Lencek & Bosker, 1999). Those beaches which today seem to be under-used relics are the great resorts of the railway age and industrial period when the aristocratic, the rich, the artisans and the relatively poor visited them in great numbers, though often segregated in their destination by their socio-economic status. The circumstances of so much seaside visitation, in the days before long-haul flight, were the expectation of taking a trip, not always for a long time, and sometimes for only a day's duration, to a seaboard destination quite close to home. Such as a trip was of course taken only by those of the masses who had enough resources to able to contemplate a holiday at all. Old seaside resorts, especially 'cold water' sites, have now for long been regarded as 'difficult' and thus they demand extra provider attention. Their range of features is often defined by their former heyday, and their attractions are large and difficult to maintain – especially so without adequate resources now arriving from tourism. They often possess, too, quantities of under-utilised accommodation, again of a style and standard of the former period of greatness.

Working toward the regeneration of traditional resorts was one key element of the United Kingdom's 1999 tourist strategy, *Tomorrow's Tourism* (DCMS, 1999). Within this framework, the emphases were on improving the quality of the amenities offered by the resorts, and directing planning toward a 'structural change in holiday trends' (ibid. p. 25). Providers and stakeholders were advised to co-operate. It was furthermore stressed that provision should be made for short breaks and other attempts at diversification in order to service new markets, such as business tourists and special niche markets. Elderly travellers were identified as one potential market for these attractions. It was decided that large grants, worth £24.4 million, should be made available by the UK government to fund regeneration attempts at seaside resorts and, according to *The Guardian* newspaper (Pietrasik, 1999: 15), it proposed designating them as 'assisted area' category to be aimed at 'the beginning of a new era of opportunity for England's coastal resorts'.

Seaside resorts have particular appeal within the definition of Dynamic Tourism, appeal that is particularly apt to attract the dynamic tourist. In Britain, for example, they have the practical appeal of being close to home markets, thus offering choices for day trips or weekend visits. There are often still public transport connections of one degree or another. They tend to offer a range of different accommodation experiences, some so traditional as to seem novel and interesting to today's travellers. The attractions offered at these older seaside resorts also probably tend to be unusual by today's criteria; they are more idiosyncratic and individualised than contemporary standardised facilities.

The beach, at any latitude and in any weather:
Deauville, France

Over and above these practicalities, however, the seaside has connotations and aspects of even more resonance and relevance to Dynamic Tourism. Where but at the seaside do we find the simplicity of an elemental, outdoor life, and intimately tied to childhood memories of a less complex, carefree existence? Many of today's potential travellers treasure sunny childhood memories of seaside visits made in an era when the resort was in a greater state of ebullience. The seaside features the properties of elementalism that appeal to the senses and spirituality in ways that correspond closely with the central aspirations of Dynamic Tourism. During the Romantic era in Europe of 1810 to 1850, these same dimensions occupied the foreground of popular taste in Northern Europe, as Lencek and Bosker remark (1999: 96).

In today's United Kingdom, the resurgence in the attraction of seaside resorts so desired by the British government is already beginning to happen, essentially through the visitor motivations outlined above. A report by Miller and Berry in the broadsheet newspaper *The Sunday Times*, states that the return to the traditional British seaside holiday, is a phenomenon tied specifically to the well-travelled middle classes, and represents 'a sea-change in upmarket British holiday-making habits that reprises the great Victorian passion for sea-bathing'. They explain the change back as this:

> The key to the revival of the British seaside is that the appeal of the summer holiday that used to be spent in the south of France or the islands of the Aegean is increasingly wearing thin for many ageing, middle-class and affluent baby-boomers. (Miller & Berry, 1999: 1.13)

There are two significant aspects to all this from the point of view of providers. This sector of the travelling public that elects to 'revisit' these ageing seaside resorts will certainly be conditioned to demand quality as a core constituent in the seaside visit experience, even while expecting simplicity. Moreover, this public will be possessed of plentiful funds to pay for its requirements and pleasures.

If this instigation is occurring, as a start, in northern, cold water, climes and the trend continues, the potential is enormous for a whole tranche of old places on seaboards that have been left high and dry and that could now undergo a renaissance. For example, we have only to list the old resorts on the English Channel/La Manche, the North Sea and the Baltic. The providers must decide, according to each individual case, how possible it is to merely reverse its appeal to an old–new audience by representing and highlighting its attractive aspect, weighed against any need to restore and upgrade existing facilities. As providers tune in to their target market, they should not assume this market to be senior citizens only. Since society in general shares the tastes identified by Dynamic Tourism, these sensibilities are not confined to the portion that is 'Third Agers'.

Gardens and parks: new prime attractions

Gardens, which appeal to the widespread desire for sensory and spiritual pleasure, are dynamic by nature. By definition, they are alive, growing and changing with the seasons. In other ways, they are also often old, whether in their physical fabric, that is, their constructed features, or in their design and concept. Some gardens, although newly configured, are seen as 'old', because the model they follow is a traditional one. Parks share these overall characteristics with gardens.

In the Parc André Citroën, which has already been discussed, Paris delivers a re-think of an old idea into a new attraction. On the other hand, the Jardin des Plantes, also in Paris, is an example of a historic old garden, but one whose most interesting elements have been refurbished. Here, the premium, aged greenhouses have undergone high-quality restoration and once again are attractive and inviting to visitors. An old children's merry-go-round, also mentioned earlier, is a dynamic feature in the Jardin and a source of great enjoyment both to the young participants and to older onlookers.

The gardens alongside the Chateau de Chaumont in the Loire valley serve as a regenerative force for an older tourist attraction; each year, the gardens are renewed in both provision and design under the auspices of the Chaumont International Festival of Gardens. This collection is now recognised as a showcase of modern, innovative garden designs. Theorists of the modern nomadic Dynamic Tourism cannot help but note that the eighth year of this Festival, in 1999, used kitchen gardens as the overall central theme, among them the portable Nomadic Vegetable Garden.

An old recreational type revisited: Jardin des Plantes, Paris, France

The former East German coal-mining region of Dessau-Wörlitz has been targeted for regeneration of a more necessary and demanding nature, with gardens as its focus. This region, which has a 25% rate of unemployment, hosts the Dessau-Wörlitz International Garden Festival, in its third year in 1999. Commenting on the restoration of the Oranienbaum Palace by the Dutch designers, Droog, Phillips (1999: 60) quotes one of the participants, Renny Ramakers, as saying that Droog was given 'The target ... to create a revival in the region through design'.

Droog delivered a range of ultra-modern, innovative products, including a line of souvenirs that evoke the orange growing of the Palace and the input of its orangery, and provide a contrast to the general greyness of the area. These souvenirs were intended as prototypes so that, as well as soliciting an increase in tourism in the area (Jongerius, 1999: 43), local artisans could adopt them and earn a living by producing them (Weiss, 1999: 17). Such items include an orange pip embedded in candy and meant to be planted by the consumer, in hopes of producing a souvenir orange tree. There is a natural 'smell tester' that allows the user to 'check the smell of plants, or other things, without causing any damage' (Guixé, 1999: 40), and a rain poncho aimed at park visitors (Jongerius, 1999: 49). A picnic mat (in spite of a ban on any such activity in the Palace Park) was designed with a map that, 'highlighted good potential picnic spots' (Phillips, 1999). Another Droog designer, Marcel Wanders, delivered 'Orange cookies' on a tray and accompanied by orange seed; when all the biscuits are eaten, the packaging serves, 'as a germinator for the ... orange tree pips' (Wanders, 1999: 55). An apple juice bottle is also meant to be recycled when empty, in this case as a birdhouse (Wanders, 1999: 52). A compact disc, 'recorded in the Oranienbaum area', delivers the sounds of 'swans, slapping doors, rain-drops, remote control Trabants, children and many other everyday sounds' (Wanders, 1999: 52). Oranienbaum is intended, 'in future ... to act as one of the gates of the historic Garden Realm of Anhalt-Dessau' (Weiss, 1999: 9).

The whole Oranienbaum garden project seems tailor-made as an illustration of Dynamic Tourism and the propensities revealed by dynamic tourists. Moreover, it demonstrates a highly unusual approach to the question of regeneration, linking an old attraction and old ways to contemporary ideas and issues.

Recognising Potential and Speciality

There are places around the world that need (or that can have if they wish) a renaissance by becoming a tourism attraction, or that require a revitalisation that tourism can help towards. These will have some existing special feature, maybe unrecognised, that once identified and exposed will show tourist pulling-power. Or they may already have an *in situ* feature that could, through good customisation and presentation, become so distinctive and appealing that it will assure a flow of tourists.

In the remote desert country of west Texas, the town of Marfa and its environs might not ordinarily be worth a journey on their own merits. However, the area also possesses a distinctive feature in the works of minimalist artist Donald Judd, which are spread across town and beyond in a permanent display by the Chinati Foundation that occupies and re-uses many buildings. The artist moved to Marfa to work in 1971, because the town and its surrounding countryside offered wonderful possibilities for art display. Now it is 'the world's largest museum devoted to the permanent display of contemporary art' (Ross, 1995: 9), and attracts a discerning group of international tourists, who come to it on pilgrimage. This example illustrates the possible reuses of existing, unremarkable resources as the basis for a new vision of any site and its eventual status as a tourism venue.

Transmuting a City: The Rotterdam Example

Our discussion has already touched on the concept of Cultural Capitals of Europe and the experience of the City of Rotterdam as it prepared to assume that title. In relation to the themes of re-birth, change and dynamic use of a city, the planned elision of that concept and that particular city is full of interest. Rotterdam is a port. During World War II, a central portion of the city was bombed. It is now seeking a new identity as a cultural venue and centre of avant-garde outlook.

The choice Cities in Scaffolding as the theme for Rotterdam's appearance as a Cultural Capital of Europe in 2001 brings a variety of different facets of the city into public display. Moreover, 'each city must have elements which will take root and last after 2001'. Several of the facets chosen for consideration in Rotterdam are related to the idea of building on a foundation of the city's past; others are related to tourism, and especially to concepts central to Dynamic Tourism. As listed in *News 2* (Summer 1999) these are:

- You and the City ('City as podium for storytellers, about the past, about how things were then, but also about how things will become');
- Rotterdam Working City;
- Rotterdam City of the Future;
- Rotterdam Vital City;
- Rotterdam Pleasure City
- Rotterdam Cosmopolitan City;
- Rotterdam Transparent City (about public space);
- Rotterdam Spiritual City;
- Rotterdam City of Senses.

Vintage Transport in New Role and Interpretation for Tourism

Old transport types and representatives are vehicles that can be reborn or enjoy an extended life by becoming tourist attractions, either wholly or in part. Steam and narrow gauge railways have served as early examples of this move in tourism; they are already veterans in this new role. Similarly, ships have been quick to adopt this new function and take on a new life, ranging from taking up tourism immediately upon retirement from their old functions, to being developed in the new role as tourist attraction hundreds of years after their original voyaging days. An example of this second group is the Tudor warship *Mary Rose*, which sank at the outset of its maiden voyage, and is now a centre of attraction on the Portsmouth harbourside.

In general, boats are very popular for recreational transport. In the instance of traditional canal boats, a great deal of the appeal is due to the waterways themselves and to the novel perspective they offer, in contrast to roads. Railroads produce this same fresh effect, as they show travellers familiar cityscapes and countryside scenes from a different perspective. In 1964, Lewis Mumford wrote during an era of cutbacks for British railways, complaining against them and maintaining:

> The kind of 'progressive' mind that is now ready to liquidate the railway is the same kind of mind that was ready a century ago to fill in the great inter-urban canal systems. Yet the only countries that still can boast a thoroughly efficient transportation network are those which, like the Netherlands, have kept their canals and have improved their railways. (Mumford, 1964: 11)

Decades later, Mumford's identification of the problem is only being acted upon in a half-hearted fashion. In the main, the full strength and

Transported by old-style transport methods: the Regents Canal, London, England

recommendation of his message seem still unrealised, unaccepted. If canals can be kept open, or repaired, re-opened and maintained, they are highly suited to boating holidays for tourists. However, up to the present point, they have been under-utilised in tourism.

Indeed, the defect of canal transport for commercial transportation, its slowness, becomes an advantage when the objective is tourism. Of course, canal boats could be brought into a multidimensional network of leisure transport and used for short-haul canal trips, where their slowness could be contrasted with other, swifter, elements in an overall travel linkage. Holidays where canal boats are the sole transportation have yet to show fully their potential to meet contemporary pleasure travel requirements. However, such holidays do demonstrate a capacity for offering engagement and participation to the tourist, while also combining several other appealing features, such as spontaneity and the opportunity to pursue serendipitous encounters. Certainly it will always offer a 'one-off' kind of experience, by virtue of the unpredictability of its activities and the chance encounters to be met along the canal route. Canal systems give the countries that possess them the potential for expanding tourism, both geographically and typologically, with all the attendant benefits such expanded tourism will bring. This possible economic contribution is enhanced by the fact that many canal systems are now under-utilised or commercially redundant, and re-encapsulating them for tourism will not detract from other, vital uses. The necessary proviso, which we have mentioned earlier, lies in the difficulty of developing redundant canals for tourism, because of the expense of time and resources needed to restore them in preparation for being used again.

The Key Requisite in Providers: Necessary Vision

Probably the most important, abiding requirement for providers who wish to redeploy resources for tourism, is the imagination to see things differently. This innovative vision is necessary, no matter what the extent of the changes involved, whether a change of function is needed or only an update within an existing role of tourism, and whether the alteration is to be applied to a single item or to a concept. In the fluid situation of Dynamic Tourism, with its broadened uses of items and places and widening of dynamic tourists' tastes, there are new opportunities for deploying old notions in new places, as well as for using new ideas in old settings. Often a new presentation of an old entity is needed. An important capacity of re-deploying, in a world of finite size and resources, is that recycling and re-energising familiar tourism assets can deliver new benefits and opportunities to destinations and their peoples, and also to tourists, in their emerging, changing needs.

Chapter 9

Staying

Remaining with Items; Or Different Motions from the Same Motives

The central premise of Dynamic Tourism is that our tourism norms have changed. Providers must now expect to plan to cater to visitors who are willing to encounter the whole spectrum of features in a holiday destination, including representative samples of daily life, rather than requiring a 'tourist selection'. In the past, tourism products were packaged, with identifiable 'trade mark' aspects, style and veneer, that separated them from those of ordinary use. Likewise, participants of tourism have not adopted the demeanour of tourism in everyday life (Boniface, 1998b: 746–9). People often joke that they never visit their own backyard attractions, unless they do so with visiting friends and relatives who are staying with them and forming the instigation. Tourism has been different, a thing set apart, on the other side of a boundary from our routine, workaday existence.

Now, Dynamic Tourism sees as the way ahead the whole world serving as a potential tourism product. So we must consider the potential in this situation for precipitating a reaction and causing retrenchment of a kind. This chapter will therefore outline reasons and strategies for 'staying'. There are various types of 'staying' and 'staying away', all with varied derivations. Visiting some places may be irresponsible, producing social and environmental damage; these considerations contribute to one kind of 'staying away'. Another impetus is recognition of the inevitable change that results when an item becomes a tourism product. A motivator could be burnout from too much tourism opportunity and travel so that people 'stay put', either through maintaining or establishing loyalty to a particular destination, or else they remain entirely at home. One more cause, and a major alteration, may be a failure of the current concept of standard tourism to attract tourists.

To resume, causes for potential tourists 'staying' might be:

- Tourists regard it as irresponsible and unsuitable to go on a visit.
- Because an item has lost its visitor appeal, because of changes that result from being 'discovered' and written into the accepted tourism roster.
- Travellers suffer from burnout, boredom, fatigue and tourism satiety – too much on offer, too much consumed.
- Tourism in its present form has lost its appeal.

When dynamic tourists respond to these circumstances by 'staying', essentially their shapes of action are two: either that of returning to 'old' products for their new-found appeal in contrast with the many options of the wider world, or that of indulging their tourist interests and impulses through a new form of tourism.

Therefore, in this chapter, 'staying' will be defined by staying true to, or returning to, one tourism product, and also by the second dimension of remaining at base. One action requires a commitment, staying committed to, or reverting to a particular existing item; the other expects an individual to stay put. The first choice can be seen as an essentially conventional aspect, while the second is more daring. However, it is important to emphasise that both ideas are forward-looking approaches, in line with the precepts of Dynamic Tourism. In this instance, staying with an existing tourism product or activity is a positive and decided action, in keeping with new criteria. Not travelling away is far from a denial of tourism altogether; it is a fresh way to interpret tourism, a new means for tourists to enjoy its benefits. In the aspects and dimension discussed here, 'staying' is an activity in which we engage as a firm act, not for the objectives of yester-year, but in order to be fluid, innovative, and contemporary. Indeed, in the context of formulating alternative versions of the tourism experience, 'staying' becomes an extremely radical act.

Many consumers see great deficiencies in the products and activities offered in the tourism market. They want to stay with the *idea* of tourism, and thus want to see tourism's run-of-the-mill features replaced. As these replacements and substitutions are made, they must tend to a format of having fewer myths in their make up and be characterised by greater veracity and integrity than before. As a result, these new offerings will deliver more fulfilment for both consumers and providers. Simply having a newly reinforced view of the good points of familiar resources will provide some dynamic adventures, as will finding fresh reasons to use the oldest and most common types of tourism products. Dynamism will also change public expectations about tourism; tourism should be so deconstructed as to have a new formulation, more suited to emerging times.

The Arrival of Maturity and Responsibility

It will be seen that 'staying', in the interpretations we have outlined, represents a strong maturity of outlook, one of the very characteristics that help to define the dynamic tourist. The individuals who lead in choosing and adopting these 'staying' attitudes participate in the obvious capacity of dynamic tourists, and perhaps also in their obligations.

The central theme of this book is that tourism has preserved for too long a simplistic, immature and static attitude, which is now out of step with contemporary times, changing tastes and the generally increased levels of sophistication in today's society. Dynamic Tourism suggests what the new methodology and outlook should be. The most accomplished and informed tourists will be the protagonists of the new approach. 'Staying' represents the most controversial, extreme dimension of the overall suggestion.

Why Stay with Old Products?

Load and exhaustion

Let us turn to the aspect of remaining with old – in the sense of long existing – tourism features and attractions. Chapter 8 examined the opportunities inherent in revisiting and recycling these tourism attractions; spreading the tourism load was one vital reason for doing so. That chapter also presented the example of the re-adoption of seaside resorts, particularly in the United Kingdom, postulating a sort of fatigue amongst affluent and seasoned tourists with engaging in far-distant travel for their holidays.

To a certain extent, this burnout motivation seems a negative reason for staying with, or returning to, old attractions. After all, many of these items are most often on a person's doorstep, since they had their first heyday before the development of long-haul travel, at a time when most tourists remained fairly close to home, both through fashion and through lack of resources. To the dynamic tourist of our day, there is a further, more deliberate reason given for staying at or near home; cutting travel distances is one way to save global resources and reduce harmful emissions by avoiding long car and plane journeys.

Staying for Good Causes

Curtailing harm

There may become a wider 'staying' reason among responsible tourists who decide, on environmental and other grounds, to stay put. Their argument might be that tourism has already brought pollution of various types – environmental, social and cultural – to the old tourism venues, and so the best idea would be for tourism to remain confined to those sites rather than spread out and pull the whole world into the tourism portfolio and zone. Considerate tourists may decide, therefore, either to stick with visiting existing sites that have already been compromised or harmed, or else choose destinations constructed specifically to accommodate all elements reasonably, rather than extending to new areas.

This attitude, of course, is based on the conventional assumption that tourism will continue as a *physical* activity, that visitors will still actually travel to venues and that they will continue to employ an existing style of operation that is like that of the past. However, we must remember that there is the other dimension of 'staying', that is, the option of remaining at home and not travelling. Slightly less extreme as an idea, but based on the same principles, is the travel plan of the tourist who embarks on only minor journeys, visiting places only within the immediate home vicinity and environs.

Co-operative care and management

A particular provider viewpoint may accompany the phenomenon of 'staying' in the sense of tourism remaining at older, developed sites: potential new destinations that need the perceived economic benefits from new tourism may be concerned at the loss of future development if 'staying away' should become a major trend. Indeed, this is one facet of the North–South global argument (Boniface, 1995: 6).

The ideal solution to keeping places unspoiled and unutilised by tourism, as it is traditionally practised, is for an enlightened global society to operate tourism in the world as a single entity geared towards the best overall global interests. One might even imagine subsidies in the form of grants given by the overall world in concert, as both tourism provider and judge, to those places that suffering from poverty because tourism has not been installed. However, to be realistic, such a Utopian objective would be unlikely to attract universal support.

While the full concept may never reach a point of entire adherence, the notion underlying such management of tourism on a worldwide basis is central to the theory of Dynamic Tourism and, specifically relevant to the 'staying' idea. The focus of the matter is for tourists to choose to manage their own tourism activity, for it to show flexibility and change. In the context of 'staying', this would mean that travellers would sometimes choose to visit an established destination, rather than a fresh one, and that they would make this choice with greater frequency than in recent years. Sometimes these choices would take the form of spending free time closer to home, with the effective results that all destinations would have 'down' or 'rest' times and no longer be over-run with visitors. Vulnerable locations would thus have some degree of respite. The outcome to be hoped for would be a middle way, a third option between the choices of 'spreading' to relieve pressure (as discussed in Chapter 6), and of entirely 'staying away' from a given place, in order to preserve another part of the globe as sacrosanct. Each of these choices is consistent with the mindset of Dynamic Tourism.

Balancing and shifting travel choice and opportunity

A further facet of this style of 'staying' is a possible shift between tourists who stay put and tourists who travel farther afield. This choice may be related to the new dynamic we have mentioned in relation to the re-emergence of seaside resorts in the United Kingdom. There is a certain logic to this shift (if indeed it is happening), and it could be regarded as part of the satiety felt by many marathon travellers, much-journeyed people who now want to stay, for a while at least, closer to home for their holidays. The opportunity of chasing receding horizons would now fall to new tourists, who have not yet explored far afield and who experience such travel as a novelty.

Intrinsic Characteristics to a Place: to Encourage Tourists to Remain or to Re-Attract Them Back

How do we encourage tourists to remain interested in existing destinations? How do we attract them back, once they have moved on? The arguments in favour of staying put are related to society. They are predicated on potential tourists who make sensible choices based on logic and directed toward such objectives as protection of societies, cultures and environments, in the interests of overall fairness to the global community. However, in addition to the causes we have mentioned, which encourage tourists to stay put and to visit already-established, familiar destinations, traditional locations may be endowed with features that *in themselves* are sufficiently special that tourists will decide either never to desert them in the first place, or that there are now good reasons for returning to them.

For example, older tourism offerings may possess innate charms or simplicity compared with today's formulations, which are likely to be complex and highly technical. Many of the sites described in Chapter 8 as worthy of renovation, such as seaside resorts, fall within this category. There can be a refreshing astringency to the simple activities of making sand castles, taking the waters of a spa, riding a merry-go-round, rowing on a boating lake, staying in an old-style holiday camp or resort hotel, and even to seeing an unreconstructedly old-fashioned museum display. Such items carry features of ongoing appeal and offer a nostalgic atmosphere and charm. They also display Dynamic Tourism characteristics and cater to some of dynamic tourists' wants and preferences. These old products can have emerging qualities of new contemporary relevance, in contrast to more recently-introduced tourist items.

Need to Travel Far?

Do we really need to travel far to find an acceptable holiday? We have already considered the suggestion that there may indeed be a transfer between those people who travel and those who do not. Were a major alteration to occur, between those countries that generate tourists and those that receive them, the implications of this reversal and the changes it would cause would indeed be varied and considerable. The tastes and inclinations noted in the theory of Dynamic Tourism are such as to encourage growth in this fledgling tendency. Already some amongst the traditional travelling sectors are staying put and holidaying at old locations near to home, at least for some of their holidays. In terms of staying within national boundaries, it should be noted that many holidaymakers of certain countries – France, for example – find little charm in journeying outside the home country when holidaymaking. Staying near home, of course, only transfers pressures from far away places to others nearby. However, when such travel is treated as an occasional 'set aside' action, meant as part of the dynamic management of the world's resources and environments, as well as permitting other people to visit high-profile sites, it does have a role in the panoply of such actions.

Done Travelling

As has been outlined already, there is another possible motivation for staying put that is also connected with self-orientation and service to individual needs. This predisposition for staying at home, rather than going afield, is not merely a case of people having travelled too much and seen too much of recent attractions, nor a recognition of the re-emerging appeal of old attractions because of their contrast with newer types of destinations. The causes of this preference for staying put, however, are very close to these foregoing cases; the strong and extreme sentiment that keeps some tourists from travelling could be *fatigue and over-familiarity with tourism itself, in its current formulation.*

To Tour or Not; and What Alternatives?

What are our alternatives, as we ponder whether or not to go away on holiday? To see why tourists, with dynamic tourists in the lead, might be staying put and rejecting tourism itself, the core character and promises of tourism must be analysed and defined. We must discover whether the products delivered by tourism providers are wanted by dynamic tourists; we could thus determine if tourism is somehow repugnant to them, and if the benefits sought through tourism are now better found by dynamic tourists in some other manner.

Tourism is essentially an occupation of our free time. It is discretionary, an indulgence we offer ourselves, and thus we make a positive choice to use our leisure time in this way. Dynamic Tourism places the responsibility for tourism choices upon the tourist. It expects an offer to the tourist of fluidity, changeability and a certain lessening of formality. The travel industry, however, is a sizeable, complex concern. This industry needs to govern and steer our choices, to see that they are structured into products of practicable provision and viability, and of a sufficiently ordered enough nature to be established and managed. The tension is obvious; needs of commodification and stability are set against requirements of variability and alteration.

If dynamic tourists are not able to find the experience they want within the system of tourism, they are likely to seek alternatives. Whereas in the past there were few exceptions available to deliver fulfilment similar to that offered by the tourism avenue, now there are other options. Given that possible alternatives are available, the ultimate act of responsibility for the dynamic tourist, therefore, if the essential wished-for elements are not forthcoming within tourism, is to reject tourism altogether and 'stay' by not going on holiday or travelling.

Staying with the Essential Components, but Finding Them in New Alternatives

Experienced tourists, dynamic tourists among them, may reach the point of feeling that they have had enough of tourism, that they are sated with the repetitiousness, in tone and flavour, of what the tourism industry provides. Since tourism offerings and sites are chosen by the industry and fashioned and promoted through it, these products, although they are distinct and different, may give the tourist a perception of a certain sameness. This perception may, of course, be a comfort to inexperienced tourists in unfamiliar territory; but the seasoned traveller, the dynamic tourist, does not need this reassurance.

Among the various products offered by the industry, the one that probably seems the most pre-eminently static and unchangeable in the consumer's mind is the beach holiday. In this case, elements of the holiday are provided for the tourist by the industry; *where* these components are delivered to the consumer, the actual location of the components of hotel, beach and so forth, may seem to be secondary considerations of little importance to the consumer. As a result, in the custom of many of the same types of products being provided by the industry, dullness and staleness may be the prevailing characteristics that register with the tourist. Therefore the strong motivation to travel may now be lacking in some seasoned tourists. So, it becomes an attractive option to stay at home and do something else to find the 'highs' that tourism might have been expected to deliver.

In his discussion of *flâneurisme*, Bauman (1994: 143) has noted, 'Play is free. It vanishes together with freedom. There is no such thing as obligatory play, play on command'. This remark can serve for much of tourism as a major industry now, since the features of fun, discovery and excitement that were present in its earlier forms have become lost. Bauman also notes the separation of play from reality, a distinction that is a key component of tourism for many people. He says:

Play may be serious, and often it is, and it is at its best when it is: but even then it is 'not for real', it is enacted 'as if' it was real, this 'as ifness' quality being precisely what sets its [*sic*] apart from 'real reality'. (ibid.)

Discussing changes to *flâneurisme*, Bauman describes it as having been:

... caught upon by a mass industry, and the cottage-industry style of catering for the need replaced by the design of high-tech marketable products and their closely administered distribution. In the process, the freedom of the *flâneur* to set, playfully, the aims and meanings of his nomadic adventures ... has been expropriated. (ibid. p. 154).

Dynamic tourists can be expected to choose to engage in individual and personally created and selected *flâneurisme* – of old demeanour. Other tourists may follow their example.

Bauman continues his depiction:

The process has come full circle: from cottage production to cottage consumption. From now on, the well-established dependency is to be reinforced and reproduced by free, voluntary efforts of dependent individuals. (ibid.)

What we see here as portrayed as occurring in regard to *flâneurisme* is what seems to be happening on the wider scale in tourism at large. Sectors of consumers, dynamic tourists at their head, will seek to reassume control of their travel experience and its style. They may reach these ends either by not travelling in the mode of the tourism industry or else by remaining at home and not journeying at all. The erstwhile tourism consumers may find their required experience through other means and sectors.

Staying with Tourism Now, or Staying Away – at Home

Dynamic tourists will look for the usual things in their tourism: escape, relaxation, excitement, stimulation and new experience. As people both accomplished in the role of tourists and with broad perspectives and purviews of interest, dynamic tourists will be looking far and wide for their satisfaction through tourism. They will be prepared to stay with or return to an old product, if logic and inclination encourage them to do so. If none of the range of choices provides them with their wants, then, on occasions or perennially, they will not bother to travel as a leisure option and will remain at base and will do something else as a substitute for the travel experience.

Tourism has been absorbing many types of recreation that were formerly not thought to be part of tourism, such as sporting and exercise activity, speciality shopping, concerts, performances, and particular events and festivals. However, even with all these inducements included on the bill of fare, with reasons of counterpoint to touring emerging and the

Enjoying the streets of home, and simple and sensory pleasure: Stockholm, Sweden

appearance of many new, appealing alternatives to travel becoming available, it is by no means a foregone conclusion that we will travel for leisure or when we do not have to. From varied and dynamic causes, our choice, in nonetheless searching to find tourism's facets, may be to stay away from tourism itself at times, or even altogether.

Our idea in this instance of 'staying' will involve looking for ways of obtaining similar sensations and the same information we would have gained by engaging in tourism. These experiences may well be found in looking more to interesting features of everyday life and to their most excitingly informative developments, to synthetic substitutes, and to technological innovations (Boniface, 1995: 79, 112). In the past, many of these potential sources of satisfaction simply did not exist.

We have discussed how *flâneurisme* in its uncommodified sense, in its meaning of play free from the constraint of others' rules of ordination and structuring, can be viewed as representing tourism as a whole, seen in its widest interpretation. On this point, Bauman remarks in his *flâneurisme* discussion:

> Today, *flâneurs* can practice the neverending play of *flâneurisme* without leaving the settee. One does not need legs to be a nomad. The seductive mountain came home, ensconsed [*sic*] in the black sheath of the videotape. You can rent or buy the tape that does the dreaming and playing for you. (Bauman, 1994: 155)

In this manner, the tourist whose taste and motivations have been expressed in the theory of Dynamic Tourism can remain in the comfort of home and have experiences similar to those provided from genuine travel. Indeed, it is quite commonplace to make virtual Internet tours, like the National Trust's offering at England's Corfe Castle. In 1999, a virtual tour of the whole city of Helsinki was in being prepared (Keegan, 1999: 2–3). The digital revolution offers a huge capacity for interaction; technology allows a participatory experience, rather than a passive one. Tours can be compiled by the individual traveller, rather than by tourism operators; it is thus possible to avoid stereotypical and commodified travel itineraries. Not only is it now possible to create substitutes to travel, but these substitutes are sometimes preferable to the real thing. These reasons may be added to those other personal and societal causes and motivations to stay home that we have already mentioned.

Before closing this chapter, we should note that a reversal has taken place with regard to tourism; what was supposed to serve as an escape, a fantasy or item of strong contrast with humdrum, routine is now widespread and ubiquitous, through industry growth and by the repeated perpetuation of the same modes of process and product. Indeed, tourism itself may not always serve as a contrasting novelty or an item of difference, because in its mechanisms, formulae and products it has become too well known, adopted, and mundane. Therefore, we may occasionally stay with more individual, less-standardised activities or, having once forsaken them, we may return to them again, as a relief. As outlined in the preceding chapters, the theory of Dynamic Tourism posits that dynamic tourists are searching wider, in terms of the type of tourism, and looking anyway for less formal tourism products. In a yet more radical decision, we could opt to avoid tourism, as such, and to seek other routes toward those sensations and rewards with which tourism once provided us, and which it could provide again, in a suitable composition.

Chapter 10

Sight Ahead

Bringing Dynamic Tourism into the Spotlight

We are beginning to define an image of Dynamic Tourism constructed along the following lines: Part 1 demonstrated the necessity for Dynamic Tourism, and then revealed indications of its emergence; Part 2 listed various processes involved in Dynamic Tourism and showed their growing relevance to tourism. The objective in the current chapter, which forms a portion of Part 3, is to refine our definition and achieve a clearer perspective from which to work toward the goals of Dynamic Tourism. This chapter will demonstrate how the features of this tourism philosophy relate to current circumstances and needs, as well as to those of the future. We will explain why these dynamic features are imperative in the overall tourism formulation, and why the format of each element should be of the identified prescription.

Change, provisionality, serendipity, breadth and lightness

Dynamic Tourism responds to an urgent need for change. Its own essence is change, and not merely as a one-off event, but on a perennial basis. This integral imperative is the most challenging dimension of Dynamic Tourism, but it is also probably its most important feature. Flexibility and fluidity join with change as emblematic elements in Dynamic Tourism. With less solidity and continuity as part of the definition, a provisional or transient element is also part of the concept. As part of an inclination towards a less structured, decided and premeditated choice of offer, serendipity appears as one governor. Another key aspect of Dynamic Tourism is breadth, that is, a broad range of applicability and extensiveness and purview. A final distinguishing feature of Dynamic Tourism is lightness. This parallels all the other concepts and the tendencies that underlie them because, along with the encouragement that Dynamic Tourism gives to expanding horizons, there is an equally important consciousness of the need to conserve fragile products and treat destination communities delicately and with sensitive respect. The dynamic tourist has a lightness of approach and treads lightly upon the Earth. The concept that we are bringing into focus, therefore, is a structure composed of these basic and fundamental features.

Mutability

As is the style of Dynamic Tourism, the very form of the concept that unites all these elements is not set; it takes its shape as soft, and malleable to a certain extent. Its emphases and predominant features differ according to specific circumstances, situations and times. The overall challenge in setting up a basic model lies in finding the right balance, remaining undefined enough to allow maximum freedom of movement, yet providing adequate structure for operational viability.

Focus on The Tourist, Changing

The tourist's inclination is the critical factor that defines Dynamic Tourism and governs its style; it is a composition focused on the tastes of contemporary tourists. We recognise that Dynamic Tourism blazes a difficult, demanding path for tourism providers and stakeholders. It requires a new attitude and focus on their part, as well as some deconstruction of established methods. It is necessary to relinquish a measure of control to the tourists themselves, giving them a greater power of choice and putting more trust in their capacities.

The overall requirement of this tenor of thought – the necessity to cater for on-going change – is the hardest test of all for providers. Along with this requirement, they will need to show flexibility to respond to the new breadth that today's tourists show in product tastes. Finding out how to achieve commercial viability and preserve a business structure in a situation where so much freedom and control rests with the tourists, will be demand of creativity by the tourism industry. This study has already offered many indications of progressive possibilities; in Chapter 11 we will give further attention to specific ways in which Dynamic Tourism can be implemented.

The point we must emphasise, when considering the tourism industry, is that it has no choice but to relate to its customers, both present and potential. If the industry grows apart from its customer base and their tastes, it will be marginalised. As this book has demonstrated (still more evidence will be offered in Chapter 11), the tourism industry is grasping the message of change and is more willing to accommodate its customers in their changing, more mature, state.

General Features in Society to Cause Tourism to Change

Since the tourist's demeanour is uppermost in conditioning the future shape of tourism, we must formulate a clear model of the contemporary human being, as well as of those aspects of society that influence the individual's nature as a tourist. Seeking out this kind of information does not mean that we forget that markets subdivide greatly, and that human tastes and capacities differ. However, it does demonstrate certain important human characteristics and offers access to people's lives, both of which appear (generally speaking) to hold considerable significance for tourism and to affect its future development.

More leisure

The tourism industry has grown tremendously, a fact that is related to the wider element of growth in the number of person-hours allocated to leisure. The essential causes of this growth lie in:

- Increasing affluence for an increasing number of people.
- Increasingly intense and demanding work activity encouraging many people to need and expect respite and contrast in leisure time.
- The tendency of professionals and other specialised workers to work longer and harder, and thus not only to need good holidays but to be better able to afford them.
- More leisure hours available because technology has taken over or reduced certain time-consuming sectors of human activity.
- Social conditioning and perceptions, reflected by legal requirements that workdays should be regulated and shortened.
- Generally longer lifespans.

Taking tourism forward and continuing its growth will require more attention to these and other factors. Increased leisure may not always translate to more tourism. Unless the developing taste of the public is met, tourism could be a bit left behind for other equivalent pursuits that offer more value and gratification.

Short attention spans and quick gratification

Our lives are less controlled by a single and formal template than they once were. We live more often according to individual style and predilection. We expect a range of options in how we live our lives, and in the choices that providers present for our consumption. With more to occupy us, with possibilities of quicker fulfilment through technology, with more diversion on offer, but with the same twenty-four hours in every day, we are also deemed to have shorter attention spans than in the past. We expect our needs to be met at once, and we demand instantaneous satisfaction. In terms of tourism, this means we want to be able to book a trip instantly and have the capacity to travel at the spur of the moment if we wish or need to.

Changes of belief and veneration

The presence of formal religious practice, which gave meaning and order to human existence, has lessened in many lives, and particularly so in those, as a sector, of the developed world and the most-travelling public. This has not so much led to a removal of religion, as to an alteration in the outlines and items of belief and veneration. Spirituality remains a constant force, and is strong in new or rediscovered areas. As regards tourism, there will probably be two main effects:

- An interest in visiting former religious sites as heritage destinations, because of consumer curiosity about things that are no longer part of ordinary life.
- The present active interest in encountering spirituality being directed towards finding the spiritual dimension in things that are not overtly religious.

Playing at opposites: complementarity

Our trend is to a majority of the population being in an urban environment and a major proportion of people's lives operating in a routine based on high technology and modern rational existence. This implies that, to a huge number of people, rural, countryside life, and all that belongs to it, delivers and represents items unknown, of novelty, and of difference. When seen in contrast with urban existence, a life bound to the land and to agriculture, although it is highly structured, is also intimately linked to features of change and chance. Elemental forces, such as the weather and the fundamentals of life cycles, play an obvious central role in country life. It can readily be seen how the countryside can offer attractions and spirituality to urban dwellers, and thus how city people can be drawn to visit the country. In return, it is equally easy to see how country dwellers can be attracted by the difference to want to experience in the city and its lifestyle.

When we see people who live in the city and holiday in the countryside (or vice versa) we are observing an illustration of a key human tendency, when engaging in the role of tourist, to choose environments and characteristics that are opposite to the ones that are found at home. This is part of tourism's function, to allow us to play at adopting alternative roles and different situations from those that are customary in our lives. We surely are meant to search for contrasting elements, for yin and yang features, with which to achieve an overall wholeness, a completeness, in our existence. This yearning is not only shown in our tendency to seek out rural settings if we normally lead city lives. It is also apparent in such inclinations as a desire for peace, simplicity and a 'naturalness' of experience on holiday if we are used to hustle, complexity and materialism in our normal, everyday life. There is an opposing motivation, for a person whose usual existence is quiet, basic, and focused on basic necessities to seek out luxury and sophistication in a 'get-away'.

Knowledge; imagination; communications

Generally, we are becoming better equipped to shape our own tourism selections through the overall improvement of education. This area still needs more attention, both through a greater effort in the formal education sector, and through the tourism industry treating us as mature consumers who are capable of using appropriate information to make responsible and sensible choices. The object is for us not merely to develop our stock of information, but also to be imaginative and creative in the choices we make on the basis of this information.

The move to become better informed can bring us closer to meeting our own changing requirements, and also to operating in an adaptable mode, with an eye towards good management and the conservation of tourism resources. It has been demonstrated that technological advances, such as the Internet and computerised systems for reservations, can aid the general process of good choice and facility management. Both consumer and provider immediately receive up-to-date information, and thus both parties are allowed adaptability and flexibility of choice and manoeuvre.

Of yet greater importance than this paradoxical appreciation of the town by the country and the country by the town, are those items that concern and interest us as contemporary world inhabitants and so condition us as tourists, too. We must always add to these factors our social conditioning and propensities to see ourselves, through the intervention of modern facilities, as part of a one-world community, first through fast information communications, then through fast transport.

Mapping tourism, globally

Though the travel activities of tourists may be seen as contributing to environmental damage, these travellers have the advantage of their first-hand experiences and observations of the world's difficulties. Dynamic tourists should have the capacity to apply their knowledge and experience to their decisions of when and where to travel, when to stay away from fragile destinations and substitute synthetic replacements for them, and when to stay at base altogether and look for alternatives to travel that give commensurate satisfaction. Their lead, now made easily by technology, ought to be towards the optimum development that is the countries of the world working together to prepare a global map of tourism and travel features.

This complex map would identify travel items, destinations and attractions across the world, showing their vulnerability and relative strengths and resistances – and to what forces. It would act as a source to allow informed choices to be made and to permit adaptations and shifts in tourism movements to be engineered as circumstances themselves alter.

Since the map would be created and maintained through computer technology, it could easily manifest the necessary dynamism. This is the objective, on the grounds of conservation and good practice. In reality, we would probably still see intervention and distortion from motivations and needs on other grounds (such as economics and politics) on the part of the providers and from the over-riding desires of tourists to visit certain spots. However, such a map, even if it were only partially constructed and used, would benefit an ultimate ideal of sustainability in tourism, and in the widest possible context.

Dynamic Tourism's Format and Influences

The main constituents

From the foregoing points, we can begin to see delineated a shape for our model of Dynamic Tourism. Indeed, the suggested tourism map could almost function as that very model. Within the Dynamic Tourism model's fuzziness and necessary propensity to form somewhat reactively, certain points can be seen as providing some structure, and serve to help the tourism industry to plan in the context of so much current fluidity. Certain key components that are present, in a context of change, provide fixed points around which much other activity emerges and revolves. These components are:

- Prevailing spirituality.
- A taste for more intangibility to products.
- Individual inclination, imagination and responsibility pertaining, though in a matrix of world-wide awareness, global consciousness, and efforts directed toward sustainability.
- General lightness of touch
- Strong input from a growing older generation's new attitudinal style and economic influence.
- Technology in service as facilitator to chosen and needed outcomes.

At various times in the course of this book, we have mentioned a contemporary tendency to emulate Buddhist principles, particularly those of Zen Buddhism. The philosophies they represent can easily be seen as corresponding to the attitudes and approaches of Dynamic Tourism. For example, Buddhist authority Christmas Humphreys outlines this way of thinking as part of Buddhism:

> Life is a bridge – build no house upon it; a river – cling not to its banks nor to either of them: a gymnasium – use it to develop the mind on the apparatus of circumstance; a journey – take it, and walk on! (Humphreys, 1990: 18)

When introducing Zen Buddhism, Daisetz Suzuki (1993), along with the two interesting comments quoted in Chapter 1 about how the desire to amass possessions is a negative aspect of personality and how the aim of Zen is to obtain a new perspective, made the following relevant remarks:

Zen ... is the spirit of all religions and philosophies. (ibid. p. 44)
Zen ... makes us live in the world as if walking in the Garden of Eden. (ibid p. 45)
Zen wishes ... to show that we live psychologically or biologically and not logically. (ibid. p. 64)
Life, according to Zen, ought to be lived as bird flies through the air or as a fish swims in the water. As soon as there are signs of elaboration, a man is doomed, he is no more a free being. You are not living as you ought to live, you are suffering under the tyranny of circumstances; you are feeling under a constraint of some sort and you lose your independence. Zen aims at preserving your vitality, your native freedom, and above all the completeness of your being.

So many elements of Dynamic Tourism may be seen as alluded to in these words. Here we see spirituality, simplicity and rejection of materialism, an emphasis on freedom, a rebuttal of static existence and an attention to lightness in travel, the assumption of personal responsibility and rejection of operating according to formulation from without, attention to depth and to understanding, and the development of fresh perspectives. So much that is understood in Dynamic Tourism is encapsulated in these lines; how and why it exists, and why it needs to exist. In effect, they could easily serve to represent its mission.

To come at these central concepts from another direction, consider the legendary past of Australian Aboriginal Dreamtime. In that time, nomads wandered through Australia, bringing its entities to life by singing. Songlines may not be seen by the physical eye, therefore, they are perhaps the intangible features that ultimately could represent this new tourism perspective. In Bruce Chatwin's novel, *The Songlines*, the character Arcady explains to Bruce, the visitor:

Aboriginals could not believe the country existed until they could see and sing it – just as, in the Dreamtime, the country had not existed until the Ancestors sang it. (Chatwin, 1988: 17)

Bruce then queries:

So the land must first exist as a concept in the mind? Then it must be sung? Only then can it be said to exist?

The discussion then continues with Arcady equating the attitude he is explaining to 'Pure Mind Buddhism'.

Emphases: fluidity, lightness, immediacy, imagination, intellect; intangibility especially

The Songline example draws focus yet again to a major feature that permeates Dynamic Tourism and its concept, and which concerns the attention delivered to spiritual, sensual and intellectual aspects. As we have already indicated, the attention of Dynamic Tourism is directed beyond the practical considerations that are the physical currency of tourism (such as aeroplanes, hotel rooms, famous buildings, visitor attractions and museum artefacts) to those aspects of the world that may not be seen, but which are nevertheless strongly experienced (such as aura of place). Dynamic Tourism includes intangible features and emphasises the importance of their current appeal to tourists. There is a generalised growth in the importance of intangibles. Noting his own occasional difficulty, as a former civil engineer, in grasping this concept, Tom Peters (1994: 13) quotes an executive acquaintance who encapsulated this idea: 'If you can touch it, it's not real'.

Dynamic Tourism's paramount concerns are fluidity, a greater sense of the spiritual and sensory worlds, and a tendency to reject the finite, the solid, the fixed. These concerns are already made manifest through a vast variety of tourism outcomes, and they will be more prominent as time goes on. These characteristics of Dynamic Tourism will be embraced and demonstrated through transport, accommodation, tourist attractions and resorts, tour operations and distribution. All these elements will be discussed in Chapter 11.

Consider for a moment the example of the territory of South Jutland in Denmark. This area, as it is recognised and described by Aage Brandt, will serve to resume the basic definition of the type of environment that can be brought to the forefront as a tourism entity by the change in contemporary tourist attitudes. This portrayal of the situation in South Jutland demonstrates a quality of understanding about tourists, their tastes and expectations, that shows the types of appreciation and sensitivities that must feature in present-day tourism. Brandt begins with a description of 'Esbjerg [which] enjoys local fame as the gloomiest town in Denmark'. In today's Esbjerg, 'Tourism is an important source of income' and 'Visitors are strangely attracted to the dreary atmosphere'. The Wadden Sea nearby, 'which has hypnotised its inhabitants down through the centuries, has a metaphysical effect on tourists as well' (1997: 128). The area scenery is 'sandscape: an abstract conversion of filtering clouds and grooves, a vast mirror waiting to be filled with light and to receive human souls and render them shapeless' (ibid p. 129). 'Feelings [here] have a restless, weightless, almost bird-like quality – in fact a tourist quality: ephemeral, bound for elsewhere or nowhere.' An equation is made between the ancient 'wading and talking' of prehistoric Cro-Magnon peoples and, 'what tourists on the coast do now'.

Brandt offers the opinion:

> Coastal rhythms – beautiful and reversible – are an infinite array of forms and non-forms. Tourists from the beginning, human beings have been waders, dreamers and, finally, less talkative settlers. The original condition was perhaps never completely forgotten. (ibid. p. 129)

In Brandt's presentation, we see tourism as a light, fluid, entity, and in this example with a product of formlessness. His depiction shows tourism as rendering spiritual and sensory dimensions, and as a matter of course showing high integrity. Importantly, it strongly suggests that tourism is a deep and elemental feature of our nature.

It is clear that if tourists are to choose what to do as tourists, in the best interests of their own fulfilment, and also in the best interests of the people and places they may visit, they will need a dose of the 'intangible' of imagination. Together with imagination, they require current, relevant information, and enough of it to meet their needs. Brochure platitudes, advertisement hyperbole and 'best face' presentations are wholly inadequate for a dynamic tourist, who is seeking to formulate personal selections from a wide range of choices. This new independence in tourism contrasts with the traditional dependent situation of the tourist, lured into buying a pre-constituted, tested product of the travel trade, already well known to the buyer in its general lines, because those lines, and the tourism offering itself, were so routine and standard.

The tourism industry needs to adopt the flavour of approach the two different companies: 3M, reported by Tom Peters as said by the former strategic planner to be 'more intellect ... less materials' and Microsoft which Peters describes as 'only asset ... human imagination'. Peters (1994: 12) says that 'the concepts embedded in these two "simple" phrases are turning the world upside down'. This same appreciation by tourism knowledge providers and the tourism industry needs to be that we becoming motivated to make our own tourism choices, rather than having a pre-set choice laid before us. Tourism providers should conform to our requirements and enable us to select and plan for ourselves, by treating us as grown up travellers and giving us the help we need 'in our adulthood' – adequate information.

This is a tough recommendation to the tourism industry, since it can be assumed to represent the loss of a certain power. However, if the tourist is 'coming of age' and demanding this change, the tourism industry in its own interests has little option but to comply. The tourism industry will maintain control of most of the basic tools that travellers need to purchase and use. Moreover, inherent in the process is a 'feel good' factor to be generated in the tourist by the very effort being made and shown by the tourism industry. All these factors should help reassure tourism providers that their industry will continue to exist in the future. However, the industry will doubtless undergo changes in its character, and its organisational make-up will probably be reformulated and its activity diversified.

We must not underestimate the importance of the fact that, in addition to a core grounding in general data, today's tourists will also need much immediate information. The Internet will play an important role in providing this fresh, up-to-date information, as well as in facilitating, and often making more direct, the process of making travel reservations. Although the traditional information media (books and educational courses), deliver the basics satisfactorily and in depth, they do not have the speed and flexibility of alteration and update needed to provide all the necessary data. It has been asserted that the Internet, newspapers and magazines can fill this function, and these are already coming to occupy a more prominent position as providers of tourism information and opinion.

Experienced voyagers, such as dynamic tourists, will have the basis of their own earlier travel about the world to help them evaluate the information they receive. Life experience provides much educational and interesting information to do with pacing and occupying the environment of the community with senses attuned, as the dynamic tourist knows well. In 1969, Marshall McLuhan noted this capacity to learn more, and more importantly, from the streets and streetwise media, than from formal education:

> The metropolis today is the classroom; the ads are its teacher. The traditional classroom is an obsolete detention home, a feudal dungeon. (McLuhan, 1970: 2).

In customarily no-nonsense fashion, Peters delivers this observation:

> It's simple ... We either get used to thinking about the subtle processes of learning and sharing knowledge in dispersed, transient networks. Or we perish. (Peters, 1994: 174)

Although this last quotation, taken in context, was meant to be general in its application, it could nevertheless be regarded as specific and suitable advice for the tourism industry and its partners, most particularly to any laggardly components of those groups.

Society and Demography

The notable independence and autonomy in choosing our travel products, activities, and overall tourism requirements, that we have identified as an integral characteristic of Dynamic Tourism, can be seen as caused by wide changes in society and in our positions as human beings and tourists. As educated individuals, informed and conditioned by our contemporary environment, we exhibit the particular characteristics, accomplishments and attitudes of our own time and circumstances.

Older and different

Our life spans are longer than those of our parents, and these longer lives have two important aspects with respect to the tourism market:

- Our longer lives offer longer potential periods to be used on tourism.
- Living longer allows the accumulation of more disposable income, more of the excess wealth that is then potentially available for the tourism market.

With so many older tourists in the population, and many individuals in that sector currently financially secure and able to spend well on tourism, there is certain to be a proportional impact on all aspects of the travel industry. The outcome of *'Grey' Market* and *'Youth' Market* studies by Maps researchers in the United Kingdom were reported by Nicholson-Lord to be that,

> The ageing of the British Population means that 'grey power' will increasingly dominate leisure and consumer spending. ... Youth and 'glamour' will become less important. (Nicholson-Lord, 1995: 5)

The study found that disposable incomes were lodged with a large ABC1 market, and also with the C2 group. Senior citizens were noted as 'mentally and physically active'. Individuals aged between 50 and 65 were identified as 'heavy spenders on leisure' (ibid.).

Another critical aspect related to our ageing population, apart from its volume and spending ability, was demonstrated in the Maps report: their outlook on life differs significantly from that of older people in former generations. We have already considered Handy's definition of Third Agers in Chapter 1. The 'greys' are having a heavy input in current changes in tourism, and also in the formulation of further needed changes. As tourists, they require treatment that acknowledges their intelligence and experience. Owing to the degree of their familiarity with the routine and ordinary sides of daily life, including routine and ordinary vacations, coupled with their new, overt capacity for daring and adventurousness, the elder members of society are looking for change from established, hackneyed forms of holidays. They seek new challenges in tourism.

Different at Any Age

Speaking of contemporary consumers in general, part of our sophistication lies in being different from old, accepted stereotypes. We do not always feel it necessary to present the same constant taste and persona to providers. The Section d retail report, *Janet and John Go Shopping* notes these complexities, adding:

> Marketers have come to realise that the old ways of labelling people are far too simplistic. Our behaviour can no longer be predicted according to our age, gender or something called class (if indeed, it ever could). Forty-somethings take up snowboarding. Girls smoke, drink, fight. Bus drivers write prize-winning novels ... This makes meaningful market segmentation far more difficult to achieve. What matters is how we feel and behave on specific occasions. We all select from our personal portfolio of alternative identities according to time and place (Abdy, 1999: 4.03)

Fluidity, flexibility, constant change and individual informed choice are all primary factors in this overview. Once again, coming from yet another angle of perspective, these particular qualities are presented in the forefront, qualities that are central to the concept of Dynamic Tourism and to the model it presents for contemporary and future tourism.

Tourism's Future Shape and Balance; and its General Continuing Viability

This book does not present tourism as a rudderless entity, nor does it recommend a tourism lacking in firm strategies, constraints and constant factors. Tourism is a complex entity and it needs certain form. As we have already mentioned, the matter requires a balance between flexibility and responsiveness on the one hand and, on the other, enough shape and weight to make activity practical. However, it is necessary to offer change, both to meet the changing needs of tourists and to accommodate the variable circumstances of tourism, with an eye to sustainability as a global society.

Caring for the Earth: a Strategy for Survival, the joint publication by the World Conservation Union, United Nations Environmental Programme and World Wide Fund for Nature, edited by Few, unequivocally presents the dilemma of tension between tourism and sustainability, saying:

> Experience shows that, to be sustainable, tourism has to be planned and regulated. It must be integrated with other land uses, especially in protected areas, and potentially damaging developments should not be permitted. Control should extend to the impact of tourism on people. Tourism's erosion of local cultures is hard to avoid. What is avoidable, however, is its occurring without people's consent, as happens all too often. Those affected by tourism, must be involved in

decisions about development and be able to modify or block what they see as inimical to their life style and environment. (Few, 1993: 73)

It seems fair to point out, as we have already done, that Dynamic Tourism's focus on lightness, flexibility and less tangible tourist products is itself a potential help to the cause of conservation. Another focus of Dynamic Tourism, of course, is simplicity; choosing the simple over the complex is a move in the direction of sustainability. The overall focus of this work, and of this chapter, on the inclinations of tourists is a deliberate choice. This is because, when providers succeed in deciphering these inclinations in a reliable way, they will have a starting point from which to find feasible ways of closely meeting tourist inclinations, ways that are also compatible with other viewpoints and necessities. Finding a balance is the important matter. Providers need to look to find *their* benefit in catering well to tourists and relating closely to their requirements. This procedure ought to reveal ways, by providing products directly or indirectly, of ensuring that there is sufficient commercial activity supplier viability.

So, What Exactly *Is* Dynamic Tourism?

Dynamic Tourism emerges from this discussion as an entity that has no set menu; its core represents:

(1) The certain guiding principles to tourism supply and operations of:
      change, flexibility, fluidity,
      lightness, informality and de-structuring,
      creativity, imagination, openness and wideness of vision,
      attention to providing plentiful enough and current information.
(2) The certain key features to tourism products of:
      simplicity,
      spirituality,
      sensuality,
      renewal and new focus,
      fresh items of introduction.

In this chapter we have sketched out a description of Dynamic Tourism, explaining it as a sound, forward-looking plan of action. Chapter 11 will consider the ways in which participants in the tourism market are moving toward the viewpoint we have outlined here as Dynamic Tourism, and the reasons they can have for adopting this as their chosen perspective.

Chapter 11

Getting There

Change in Process

The concept of Dynamic Tourism has been formed from observations of changes in tourists and society. These changes are either already emerging, or else may be seen as set to occur. The Dynamic Tourism concept is a recommendation of alterations that need to be made in tourism as it is lived, for the good of tourists themselves, the industry, the stakeholders, and society as a whole. Tourism, in the format defined by today's industry, seems too much to be a perpetration linked to tourists as novices; this dated approach makes the provisions of Dynamic Tourism even more important and necessary than they first appear. The market must make certain that sensitive, subtle and varied provider approaches are now the style and norm, and that these approaches are attuned to today's society and to the maturity and experience of today's tourists.

Even though tourists are in favour of change and are already showing an alteration in their travel actions, further adaptation may place considerable demands on them. For most of the tourism industry, adopting the alterations outlined in this book is a challenge of major proportions.

Causes for Change Noted

Chapter 2 dealt directly with the signs of change in contemporary tourism, a theme developed further in subsequent chapters. This final chapter will outline how the tourism industry, along with its partners and stakeholders, is meeting some of these new imperatives. It will also make recommendations, and demonstrate what is still needed for the outcome to be entirely within the definition of Dynamic Tourism.

The WTO has published a discussion paper based on the results of a study commissioned in 1990 with the firm Clevedon Steer of the United Kingdom. This paper, *Tourism to the Year 2000: Qualitative Aspects Affecting Global Growth*, describes various changes that will probably affect tourism. Among the major changes noted in the study are these:

> More *flexibility in working patterns* will in turn lead to more variation in holiday travel demand, e.g. growing demand for different types of destinations and activities, more tourist business outside the conventional peak seasons.

Improved educational levels (and particularly higher education) are increasing different people's awareness and knowledge of each other ... [the] globalisation process has many impacts on, and implications for, travel and tourism. The most fundamental of these is the fact that increased travel is both a reason for, and result of, the global life style. The resultant growth of interest in other societies fuels the desire to travel and to seek varied experiences from such travel. Consumers are demanding new, more imaginative and varied tourism products and services. *Tailor-made travel arrangements will grow at a faster pace than pre-packaged holidays over the next decade.* (WTO, 1991: 10)

Several of the key aspects of Dynamic Tourism, as defined in the present book, were already to be seen in the *Tourism to the Year 2000* study. The need for flexibility, the demand from tourism consumers for more variety and imagination in the products offered to them, and the increase of knowledge and awareness in these same consumers, these are all points touched upon in the preceding chapters. Enough time has elapsed since 1990, when the Clevedon Steer discussion paper identified these themes, for the tourism industry to respond to the study and implement its suggestions. At least, there would have been time enough for such a response, if a dynamic and reactive climate had been present. Outside observers may perceive the tourism industry as too slow and inadequate in its reaction, with suppliers neither sufficiently attuned and attentive to their market nor adequately radical in their response to change.

Actions and Activities

Making optimum use of information technology?

In September 1996, the newly appointed Chairman of the British Tourist Authority and English Tourist Board, David Quarmby, used his Keynote Address to an international conference at the University of Northumbria to make a stringent observation, in reference to the tourism industry and the information technology at its service:

It is a classic application of information technology to collect, collate, interpret and present information about tourism products which is tailored to the interests and aspirations of thousands of distinct market segments. Yet my first impression ... is that, except for air travel, the information revolution has largely passed this industry by.

Over-integrated?

One seemingly contradictory development, at least if we acknowledge the necessity of catering to ever more fragmented, individual and identifiably different tastes, is the incorporation and take-over of small or niche market operators. These tourism providers are notable for their closeness to a specialised customer group and its particular preferences. One example of such a take-over, lamented in a 1999 article by Roger Bray, was the disappearance into Thomson's of the small unconventional operator Simply Travel Holidays, as well as Headwater (a provider of bicycling and walking holidays of discernment). From the point of view of the operators who absorb the small providers, the essential requirement is that customers should perceive no negative difference in the products or delivery of services of the firms that have been taken over, when compared with those of small providers that have yet not been taken over.

Noel Josephenides, managing director of Sunvil and a member of the Association of Independent Tour Operators, feels that the acquisition of a small firm by a larger one will probably change it, as well as causing a loss to the customer of detailed attention. He believes that yet another result of such take-overs is that, since large travel companies will often own items such as airlines, there could be 'unfair competition, which could erode consumer choice' (Bray, 1999: 8).

Break-ups and shifts

Apparently, the third largest tour operator in the United Kingdom, the Thomas Cook group, has carefully noted consumer inclinations to choose for themselves. In September 1999 it launched a company called JMC, to show 'a new attitude'. In an apparent deconstruction of an old entity, JMC paraded the slogan, 'Unwrapping the package holiday piece-by-piece'.

The heartening aspects to what has happened to Headwater and Simply Travel Holidays and in the establishment of JMC is that the examples reveal that new consumer tastes have been recognised and seen as sufficiently mainstream to justify the attention of mass market operators.

Meanwhile, Thomson has offered the 'Just' holiday brochure, described by Balmer (1999a: 23) as 'the package ... cut back to basics'. According to Balmer, Thomson says that it is aimed at 'experienced, independent-minded travellers who don't want much support'.

Environmental protection and cultural care remedies

The travel industry has been making an effort to demonstrate that it is aware of issues and responsible in its attitude. Green Globe was mentioned in Chapter 3 as a WTTC environmental programme. Amongst its members are many of the major transport, accommodation, and tour operators, and quite a few Tourist Boards also. One member of this group is British Airways. We can see a sign of individual member effort in an initiative taken by the package tour arm of British Airways, British Airways Holidays. The initiative takes the form of a booklet of travel advice aimed at air customers, with suggestions on environmentally responsible travel and

Cultural Tips. However, while demonstrating the good practice of listening to customers, the booklet somewhat taints its impression of high-minded responsiveness and environmental responsibility. Travellers are advised of the option to inform their local British Airways representative of any environmental concerns connected with their destination while on holiday, but the twin other option suggested is for these to be directed to the Marketing Department – a fairly crude revelation of the core motivation of the general initiative behind the booklet.

Potential threats to the environment have become such strong emphasis topics in the past few years, that it would be a very unaware individual indeed who was not conversant with a variety of such hazards from various sources, including travel itself. These threats must be of special concern to members of the tourism industry. The degree to which tourism providers will act upon common environmental worries in specific circumstances will depend on whether a product is likely to lose its viability if no action is taken. Providers are more likely to act if a particular group of customers seems likely to care enough about the environmental threat to reject the product, when there is no sign of the provider improving circumstances in order to reduce or remove the danger.

The dynamic tourists, as we have defined them, may be expected to be among those committed individuals who are truly concerned about the environment. Therefore, those sectors of the travel industry that cater to this market segment must notice and respond to their stance. The example of the Grotte Chauvet, which was discussed in Chapter 2, illustrates the contemporary change in society's attitudes about environmental concerns and one particular response to the change. In this case, a cave decorated with prehistoric paintings has been kept out of general public view from the outset of its discovery in order to protect the ancient art from damage. In contrast, the Lascaux cave art was nearly visited to death before suitable intervention occurred.

Following such a course of action as removal of a precious artefact from the public is an easier option, of course, in an independent situation, where there is no pre-existing commercial prerogative to be satisfied by tourism. However, the case of the Grotte Chauvet shows how operators of comparable attractions endowed with special, fragile features may need to deal with exhibiting those attractions in the future. Diversification of some sort will be necessary in order to generate revenue. Perhaps a replica of the fragile attraction can be provided, allowing visitors to interact with the substitute site, without harming the original. The decision to charge admission or not can be taken according to the particular circumstances and how much overall income is required. Such basic decisions will shape further choices about whether or not side-attractions should be considered, whether they will be defined as revenue-generators themselves and, in relation to the main feature, whether these possible secondary attractions should be seen as central or supporting income earners.

Midgley cites an example of efforts directed towards preservation and appropriate development of the much-visited Yorkshire Dales of the United Kingdom. A no-helmet, virtual reality tour of walks and climbs has been imagined, together with the possibility of an aircraft or hot-air balloon experience of the Dales. Such a virtual reality tour permits a potential whole-world spread of customer types, and does not exclude disabled individuals (Midgley, 1996: 7).

Sometimes an activity that seems to be friendly to the environment, and is even promoted as such, can come under attack as environmentally harmful. One growing trend encouraged by contemporary tourism, for example, is enjoying cycling and horse riding in natural surroundings. Surely these can be seen as benign activities? However, restrictions have been placed on just such activities in the New Forest of the United Kingdom, which is said to be 'reeling under the impact of tourism' (McKie, 1996: 4). Plans for tolls and parking fees for cars are being considered, but there are already restrictions on horse riding, and the use of some gravel tracks of a bicycle network has been banned.

Attempts here at discouraging certain tourist groups seem to be focused on day-trippers. In contrast, overnight tourists are wanted, since they deliver a much greater proportional income to the area. The New Forest is in Southern England; a much-populated area where the potential for a large numbers of day-trippers is huge, while the corresponding income to be raised from their visits is currently small. However, these same unprofitable visitors can disproportionately increase the unwanted aspects of tourism through erosion of the environment and the special aura of the New Forest. These effects are familiar in many situations.

Other solutions to such difficulties as those in the New Forest, which allow for wider use than only that of day visitors, are these:

- Make appealing alternative attractions available in places where little or no harm can be done.
- Offer other features to provide revenue, so that lessened visitation or a full or partial ban on fragile sites may be imposed.
- Introduce taxation at vulnerable places in order to induce traffic away from them.
- In cases where attendance is insisted upon, when there is resulting damage, the people who cause the damage should be required to pay for its repair.

Chapter 3 reported on the example of the Peak Environment Fund and the work that is done under its name, i.e. informing and engaging visitors as partners in maintaining the environment. This organisation is also engaged in a project with the National Trust, Lake District Tourism, the Conservation Partnership and local hotels in four participating areas of vulnerable environments and with heavily visited sectors. Along with the Peak District itself are England's South West, its Lake District and the Snowdonia area of Wales. The project involved the addition to each hotel room charge, for whatever length of stay, of a '£1 passive donation ... unless the guest requests otherwise' (National Trust, 1999a: 19). Thus, visitors render direct financial support to the care of places that they choose to visit and whose wear and tear they contribute to.

What Does a Dynamic Tourism Industry Need To Do?

Certain general points must be made in relation to the types of approaches and products offered by tourism providers. In accordance with the formulation of Dynamic Tourism we must begin by recognising the changing preferences of tourists, and the basic necessity of catering to those changing tastes. This close attention to the tastes and concerns of tourists and society as a whole is the first essential point; providers must never lose sight of the moves and changes in the overall audience and in their customer base.

There is a general consumer trend towards greater environmental concern – at least to the public articulation of such a concern, and also, to some extent, to the adoption of a lifestyle consistent with this concern. This trend requires thoughtful reactions from the tourism industry. Today's tourism is being re-defined to encompass a much wider portion of life in general, and to experience that wider portion in greater depth, with the aspiration of achieving overall wholeness and balance. This broadening and deepening of tourism must be taken into account while characterising the contemporary tourism offer.

Yet again, tourism providers must consider the new tendency of travellers to prefer less formality in things, while remembering that this tendency does not mean that tourists will relinquish structured products and approaches and commodified items altogether. Still other new emphases for providers that may be perceived through study of the contemporary tourism market are:

- the need to reflect the imaginativeness, creativity and flexibility of the individual;
- the imperative need to notice and acknowledge the increased education and knowledge of today's travellers.

What are the practical outcomes of these considerations? In general terms, we may expect to see the following effects, with the particular results spread throughout the various tourism sectors.

Transport: review and reinterpret

In transport, air travel is already showing polarisation between business-class and tourist-class travel. In business class, there is luxury, convenience and efficiency, with built-in extras calculated to ease the speed and frenetic pace of business life. In tourist class, the product offer is more basic and simple, aimed to suit casual travellers. The tourist-class passengers may have put together their own travel packages. They may have made the conscious decision to allocate fewer funds to their plane travel than to other elements of their trip.

Under the precepts of the theory we have called Dynamic Tourism, rail travel is due for a revival, because of its environmental friendliness, as contrasted with travel by plane and private car. Rail travel also offers frequency and the delivery of a wide range of destinations, as well as the sheer quantity of travel opportunities and their immediate availability, compared with air travel. Trains have another advantage over planes: they deliver their passengers into stations situated within city centres, near established hotels, restaurants and entertainment and business sites. Since trains travel at ground level, they also can stop to pick up and deliver travellers at many more stops along a given route from starting point to final destination.

The case of train travel, of course, evokes one general dilemma for dynamic tourists: the tension of choosing between public and personal means of transportation. Public modes, such as trains and buses, are more environmentally friendly, more responsible in use of fuel and other resources. Private conveyances, on the other hand, particularly private cars, can be judged as less kind to the environment when their total usage is great, and represents an extra use of fuel and terrain pressure, when compared to public transport. The outcome of this particular controversy is likely to parallel the outcome of changes in tourism as a whole. The present trend is toward less routine tourism selection, with more choices being made on an individual basis, calculated to suit best all the circumstances and personal requirements defined in one individual moment. One viable choice may be packages of transport, where the long-distance travel component is completed on public transport, with other destinations subsequently reached on foot, by bicycle or by some other individualised means. We must note that public transport facilities should make better allowance for travelling with a bicycle, and make a much greater effort to make it easy for bicycles to be carried along on public vehicles.

Accommodation: separating and polarising

Like the transport sector, the accommodation sector is most likely to display a separation between catering to needs for the luxury and for the basic. At the basic end of the spectrum, inexpensive and well equipped but 'no frills' hotels are already burgeoning. Hotels have always enjoyed the distinction of contrasting with the everyday, whether through their luxurious or otherwise 'different from home' components. They represent opportunities for living in a contrasted or heightened way, in relation to normal home life. The vastly different examples of the designer hotel, set against the 'pseudo-poor' huts affected by exotic resort hotels, show the contemporary taste for rooms and facilities that stand out from the ubiquitous bland and boring type of hotel style found the world over.

In the future, the search for differentiation should extend particularly 'downward': more interesting accommodations should be offered of a simple alternative style and interest but that would nonetheless provide excellent quality basic facilities. The tent, for example, has already been redesigned out of all recognition to be at once simple and portable and easily erected. However, even more creative versions of the basic type could be offered, while massive and complex examples of tent accommodation also exist.

In contrast with the way the tent has evolved as tourist accommodation, the caravan seems locked in a 1950s style of spurious glamour, with little evidence of a contemporary connection. Of course, this very fact could lend it a certain charm in the eyes of the adventurous, interested dynamic tourist. In general, we may say that today's technology and creativity are under-used resources in the effort to deliver new sorts of lodgings that are efficient, energy friendly and inviting.

Destinations and attractions: diversify, shift, and deploy data

Destinations and attractions are the locations where providers tackle the hardest, most complex tasks of catering to dynamic tourists and obtaining revenue from the circumstance. In the past, long lead times were required in order to prepare these products for visitors, which lead to the perception that such sites cannot alter enough to be flexible and responsive and dynamic overall. Reality in these cases is far, far less extreme. Essentially, the requirements of Dynamic Tourism in this situation are two-fold: (a) novelties and variations must be delivered to the customer; (b) the provider must reap sufficient income to maintain features and receive profit or earn attention.

A third demand is added for destinations that are, or contain, a precious, vulnerable feature or environment. Unless that destination is very small indeed, it should possess both a range of possibilities and room for manoeuvre. Therefore, the destination has elements to be marketed to different groups in different seasons. Some features can be visited more than others, and over the course of a given time span. The traffic flow to attractions within the destination area can be managed and implemented by ticketing, by varied pricing, and by well-focused promotion. Through a good acquaintance with visitors and their likely preferences, providers can cultivate the illusion of presenting their customers with free choices, while carefully manipulating them in their decisions.

Part of that manipulation, discussed in Chapter 10, is in what, if any, admission price, is required from the visitor. Where an entrance fee is not appropriate or is decided against, it is left to the provider to raise income from add-on attractions and facilities. In order for providers to extract funds from visitors in this way, they must know their visitors well enough to provide what will attract them and also to produce an occasion that will encourage their spending. The correct timing and environment for the offer, coupled with the delivery of a product that is unique, associated with the particular attraction, ideally integral to the site and unobtainable elsewhere, are features that encourage visitors to purchase.

As noted in Chapter 3, till receipts can yield much information about consumers, including the main times of visits, goods purchased and time of purchase. This information is useful in managing a resource and its staff in an efficient and cost-effective manner. Pre-bookings and reservations also help in this general process. The Disney theme parks, with their mass of shops and food outlets, are exemplary illustrations of the fact that income from a venue is far from limited to admission fees and the main attraction.

A different example of such collateral money-making is a museum where the management grasps how much potential income lies in opening a café or distinctive restaurant, or in marketing excellent goods through a museum shop, targeted to appeal to its specific markets. The Geffrye Museum in London is one example of this category. The Geffrye's original base of old almshouses has been expanded by the recent addition of an extension, designed by the flamboyant architect Nigel Coates. Intended to present English domestic arts, the museum delivers distinctively English food and drink in its high quality restaurant. It has a noteworthy shop, augmented by a Crafts Centre that both exhibits and acts as a sales agent for work by the many designers who live in its immediate area.

A similar case in point is the Tate Gallery at St Ives, as we have already mentioned in Chapter 6. This houses an excellent café-restaurant and a distinguished bookshop. Indeed the art tomes displayed there are often forbiddingly expensive, but price is no deterrent to many members of the Gallery's cultured and affluent audience, because they appreciate the convenience of finding such desirable and readily available works. With a high-spending customer base, providers need not make a large number of visitors their first priority; it is sufficient to bring in enough of the target audience to yield the target amount of expenditure. One potential bonus for providers in having a smaller number of visitors is a reduction in the damage to the physical fabric of their attraction. Better preservation of the feature as a whole should be the useful result of a smaller, more cultured and affluent visiting public.

Dynamic Tourism predicts an emerging propensity for visitors to adhere less to formal attractions, which would lead to a loss of income from such attractions. Dynamic Tourism also directs that tourism should be spread more widely, and thus away from well-beaten tourism paths, again leading to a potential loss of revenue-production there. With intelligent, directed effort, many new ideas and/or alliances of income generation could be developed to respond to these shifts away from traditional focus points.

In situations where open sites, such as villages, towns and cities, are unbounded and not in single ownership, provider co-operation or majority rule may govern the choices made to direct and profit from tourist movement. As a result, there may ultimately be a need to impose legal restraints to make necessary changes on the general behalf. Such a development will necessarily call for a system of grants to be given from taxes to individual sites that lost revenue because of a decision intended to benefit the community as a whole.

There is one final point to make in relation to tourism providers needing to generate income while today's tourists have the freedom of more choice, wider opportunities and generally less limited situations. We must draw attention to the way that the tourism industry currently overlooks many revenue-generating opportunities for providing add-ons, spin-offs and special associated goods. Other sectors of the leisure industry now routinely market goods in relation to movies, special art exhibitions, sports teams, and so forth, as well as carrying themselves forward into the general purchasing scene. To serve as souvenirs and to generate revenue, such items need not be available only on site; mail order or Internet sales vastly increase the potential capacity for off-site income of a distinctive, 'brand name' attraction.

Of all segments of the tourist industry, destinations and attractions are potentially the most liable to need adjustment and complex arrangements. These arrangements must always reflect the changing tastes of the contemporary tourist market; adjustments must maintain site existence, viability and relevance (and keep that relevance in the public eye). Moreover, most sites must earn revenue. Even the most detailed information available on the whims of tourist tastes cannot make it easy for providers to deliver exactly the right products to cater to those tastes, with viability, and to judge how long a prevailing tendency will last. It is difficult to gauge whether a sign of inclination will turn into a strong market trend, or whether a current trend will survive or represent only a 'passing fancy' not worth the expense of changing for.

Following the solar eclipse in Britain and other countries in 1999, there was an immediate niche market impetus for the relevant travel trade representatives, Explorer Tours 'which specialises in astronomy tours' and Sunvil Discovery. This was the sudden need for readiness to respond to inquiries about travel and facilities for the next total eclipse in 2001, over Africa (Balmer, 1999b: 6). Having experienced the 1999 eclipse phenomenon, should market providers have anticipated the follow-up interest?

In Nepal, tourism developers have apparently 'clocked' the new interest in Buddhism. They plan to deliver a complex that will accommodate '1,000 tourists a day' at Lumbini, 'one of the four holiest places of Buddhism'. Along with possibly as many as a dozen hotels, this complex will contain museums, a 'Lord Buddha' wildlife sanctuary and a sculpture park. Plans also include an accompanying new transport infrastructure of roads and an airport. There is tension between the tourism boosters, who focus on the potential of increased tourism income for an economically strapped nation, and the spiritual community represented by some Nepalese monks who fear a loss of aura.

Indeed, the input, participation and responsibility for this project extend outside the borders of Nepal, throughout the wider community of Buddhism. Guesthouses to accommodate pilgrims are being built by many of the Buddhist countries around Nepal, and much of the necessary funding for the project has come from Sri Lanka and Thailand. India is funding a large temple with a library. A park is being prepared to house the anticipated arrival of Buddha statues from other Buddhist countries. A subject apparently of most concern is the issue of whether an entrance fee can be charged at such a spiritual site. One Nepalese monk is quoted by Burke (1999: 23) as opposing, 'any commercialism [that] "would be very much against the spirit of the site"'.

Now, it is indisputable that a huge financial investment has been made to develop this centre for tourism, and that there are justifiable, insistent reasons for the development to be commercial. Keeping in mind the commitments made to the project, and the large, varied financial investment, we must still ask whether the idea is 'right' in conceptual terms. Have the powers-that-be judged correctly that there will be sufficient motivation, for sufficient numbers of tourists, to justify seeing the project in this way? Yet another related question: given that a search for spirituality lies behind many potential visits to the site, might an inclination for spirituality predispose those potential visitors to prefer another outcome? Might not more visitors and/or income eventually be obtained by providing a different style of destination, or by displacing commercial venues to another associated, but separate, place? The final question may be whether the overall concept steers a little too closely towards the mythical Club Zen, created by Douglas Coupland in his book *Microserfs* (Coupland, 1995) already mentioned in Chapter 2.

Getting There
175

Keeping watch: on the tourist, and society

As the industry learns how to base judgements on close attention to visitor opinions and intentions, it is useful to consider the current predicament of the National Trust. It was noted that visits to National Trust properties where admission is charged were lower in 1998 than in 1997 (National Trust, 1999b: 13). Country houses make up a major part of these properties. Were these grand old houses losing their intrinsic appeal to the public? Had these destinations retained their basic attraction, but fallen out of tune with contemporary visitors through some lapse in their style of presentation?

Truth to tell, a reduction in numbers of visitors reduces wear and tear on a site, thus helping preserve it for the future, which is the main mission of the National Trust. Yet such a reduction also means less revenue and less chance to educate and win public goodwill. The dilemma was a considerable one. Was it important to demonstrate the relevance of these properties to contemporary visitors, to make adjustments, and bring visitors back in greater numbers? Was the need to generate income targeted for preservation a more important imperative than passive preservation through fewer visitors? Should attention be directed toward optimum methods of income generation (souvenir shops and tearooms) to encourage visitors to leave more money behind them? Catherine Bennett, a journalist at *The Guardian*, gave her general opinion of society's tastes:

> Today, to judge by the National Trust's declining attendance, the scattering of aristocratic spoor – a stray *Country Life* here, a greasy drinks trolley there – no longer has the same magic. (Bennett, 1999: 9)

One interpretation, in the light of society's more democratic impulses (which we might note as an implicit aspect of Dynamic Tourism and the public that Dynamic Tourism serves), is that the main underlying difficulty faced by the Trust is fundamental (Boniface, 1996: 110).

The decline in visitor numbers to country houses can be due to dull, dated and 'fixed' presentations, but it may also perhaps be due to deep-seated failure in the motivation of visitors to stand in awe before scenes of aristocratic lifestyles. Maybe there should be a new concentration on the more dynamic features of these properties, the gardens, for example. Such a focus would emphasise the characteristics of spirituality and peace to be found in gardens, which we have already discussed in earlier chapters. A focus on gardens would also appeal to a wide sector of visitors, since so many people grow plants and enjoy the activity.

We have already referred to the Canadian Museum of Civilization as an example of an attraction that shows attitudes suited to the times. Fortunately, the Museum's Director and a member of staff have described the details of its formulation in a book. This book is a good example of the attitude and calibre of thought needed to deliver an attraction fitted to contemporary times and interests. In Chapter 1 we mentioned the humanism overtly shown in the design of the Museum. MacDonald and Alsford clearly explain the necessary 'people orientation' and democratisation of their vision, saying:

> Cardinal [the architect] was instructed to ensure facilities and amenities catered to the safety, comfort, and pleasure of visitors. (MacDonald & Alsford, 1989: 18)

Discussing democratisation in relation to museums they explain:

> There are many aspects. ... One is to make museums, their collections, and their information resources more accessible physically, intellectually, and emotionally. (ibid. p. 40)

These authors understand the position of museums as part of the competitive situation in the leisure and culture industries, presenting their institutions to changing audiences:

> Expectations of museum visitors and the tourist population generally are rising, partly because of other cultural or recreational institutions with which museums are in competition. Visitors continue to seek both educational and quasi-religious experiences in museums, but they also want to be entertained, to have their senses stimulated, and to be offered comforts and conveniences not previously found in museums. (ibid. p. 45)

MacDonald and Alsford discuss accessibility as a 'cornerstone', with its various dimensions applied to the Museum and reflected in its policies. One aspect of their definition is 'communicating in terms the visitor can readily understand, as well as being willing to listen to them in return' (ibid. p. 65). They summarise their approach as, 'providing a range of expectations to cater to every need, every predilection. It is *sharing'*.

Deconstruction

As we have moved through the chapters of this book, the succeeding arguments and examples have served to describe current practices and to indicate how tour operators and travel agents will need to think and adapt as they provide services aimed at filling the prescriptions of Dynamic Tourism. While taking countenance of and portraying many contemporary attitudes will be their objective, perhaps the best way for these tourism providers to proceed is to engage in 'deconstruction'.

An example of the style needed to deconstruct propositions in the tourism market is that of the new firm JMC who, as already described, have taken apart the tourist 'package'. The essential requirement is for suppliers to display basic units of travel possibilities to potential tourists, so that they can put those elements together and add to them in various permutations. Of course, the basic structure should be as related and adaptable to the changing tastes of the target market as possible. Contemporary tourists often wish to make more of their own choices than ever before, and to have greater flexibility of selection. This fundamental change in the market produces the context for the new provider method. This format is a reasonable compromise; it allows providers to avoid relinquishing control to the point where costs escalate dramatically for all concerned, and also to avoid becoming too complex for operational viability.

It should be emphasised as well that part of the increased range of choice for tourists will be the option to select a package vacation tour in more or less the old-style format, with all the main elements of the trip included as chosen by the operators. The ease and efficiency of the fixed-price travel package make this option attractive, even for Dynamic Tourists, especially when the complete price is less expensive than the same elements would amount to if purchased separately.

Distribution: new avenues, and the travel agent to develop a special niche

As changes come about in the system of distribution in the travel market travel agents, simply to stay in existence, need to offer a superior facility of advice, service and economy of effort for their clients. More outlets for making travel enquiries and arrangements must be available, in general, and a generally greater range of tourism options must be offered. Travel agents must therefore be prepared to present an obvious improvement in the quality in their services; they must be ready to compete with more new ways to purchase travel options, with travel research and sales on the Internet among the methods and mechanisms of direct sales and easy purchase.

Tourists and Industry, Together

Information and intelligence

Supporting information will be of paramount importance in the emerging, relatively uncharted, and more demanding circumstances of the travel market; this information will be equally necessary for tourists and their travel providers. This book has sought to emphasise that particular point. Both consumers and suppliers will need more than information, they will need intelligence. As Handy (1994: 24) remarks, 'The new source of wealth in our societies is the intelligence quotient'. He notes that intelligence is likely to go where intelligence is, and warns that a society will emerge that is divided between individuals with greater and lesser education, 'unless we can transform the whole of society into a permanent learning culture where everyone pursues a higher intelligence quotient' (ibid. p. 25).

The contemporary need for continuous learning is well recognised. One example of this recognition is the National Trust's Minerva scheme, which was mentioned in Chapter 3. Those who choose not to pursue continuous learning run the risk of being at a disadvantage, as Handy warns. Not all learning is acquired through formal education, of course. As MacDonald and Alsford maintain (1989: 41), 'Museums can provide individualised and relatively unstructured learning processes at far lower charges than university fees'. Thus the ancient institution of the museum preserves a strong contemporary function, as an educational resource as well as a tourist destination.

It may be argued that today's dynamic tourists, who are defined as generally sophisticated, experienced and intelligent, can serve some subtle and unofficial role as distributors of new, wider information to their hosts as they travel and meet people in less developed countries. It is true, of course, that many travellers do so expressly to learn from other peoples and cultures; so, while the tourists are teaching their hosts, the hosts are also likely to be actively engaged in the reciprocal agenda of educating their visitors.

Training in context and understanding, not merely in immediate skills

Dynamic tourists should be observant, attuned to their surroundings, and sufficiently informed to be able to choose their travel activities and destinations well. In the Dynamic Tourism context, the tourism industry is required to study and refine what should be its philosophy of suitability, and to learn how to remain alert for hints of change and their implications for the future. As a first point of departure, tourism industry providers must develop a strong, consistent understanding of consumers and their inclinations. Industry personnel must come to understand and value how these two elements, philosophy and consumer understanding, must be foundation stones of their business. With this dual foundation in place, there will be a context for the making of practical decisions.

When providers learn skills and accomplishments only by doing them, they find out only about operating activities and outcomes. Thus, they receive training only in an ensuing aspect. There is a risk that training does not deliver to the provider the whole mindset of operation, or else not of sufficient quality, and instead concentrates on rendering practical and vocational aspects of tourism, in business terms. The International Institute of Tourism (founded in 1988 by George Washington University and the WTO) seems to be in a privileged position to give a lead in influencing international tourism policy. These intents are accomplished there through teaching, research and continuing education. Green Globe (1996: 4), a member organisation of the WTTC, acts toward sustainable tourism in various ways. One of these ways is its active training programme, developed with an understanding of the two distinct strains in this 'the general thrust of creating and stimulating interest and knowledge, and the more specific training for individual members of staff in issue-related topics'.

The Summary and Essence of Dynamic Tourism

Tourism demands change and is itself in the process of change. Knowledge about these changes (how tourism is altering and how it needs to change further still) is a vital aspect of the process of transformation. The definition of Dynamic Tourism developed here identifies an emerging personality for the tourist that is only gradually being noticed by the tourism industry. The change in tourism is derived from the consumer, the dynamic tourist, and that person is represented both as a traveller and as a caring world citizen. Fluidity, flexibility, lightness and breadth are the main attributes of Dynamic Tourism: they signify its theme of alteration. In order for these components to feature and to be effective factors, knowledge, imagination, sensitivity and technology must also be present. Within the Dynamic Tourism compendium must be incorporated high technology, solidity and spirituality, intangibility and sensory appeal, and complexity coupled with simplicity.

In Chapter 1 we cited Plog's (1994: 50) call for common aims for tourism, for its own benefit and for the good of the world. Concerted action for tourism is the essential. Within Dynamic Tourism, however, is the implication that our society does give the correct weight to its interpretation of our tourism, even as we move toward the desired general effort. In order to progress with a good overall effect for tourism, we need to understand what tourism is and how it is developing. We must not limit ourselves to seeing it only in portions, with some of those portions already outdated or redundant. Dynamic Tourism's chief suggestion is that we do not see our tourism in its full, balanced image. We see too much that is obvious, superficial, hackneyed and materialist, in contrast to witnessing change and seeing how matters related to nature, vision, imagination and our own profound being feature as a part of tourism too. There has been too much emphasis on seeing our own perpetual expectations, rather than on noticing and characterising all that is now required as part of tourism. We have leaned too heavily on the history of what we have been as tourists, and not moved toward what we are in the present and what we could be in the future.

An encapsulation of a tourist who is reflecting Dynamic Tourism is of a modern day pilgrim and voyager, who is setting off into the world, equipped with knowledge and understanding, and who is carrying open-ness to experience. Physical luggage may be reduced to a small, efficiently equipped knapsack, easily carried on the back. Clothing is light, spare and protectively efficient. Feet are shod as lightly as the latest and best modern technology can offer. Dynamic Tourism is this. It is an attitude that must be adopted by the tourism industry in order to pursue its own venues of interests. It is a way to offer choices and flexibility to tourists, in order to get them where they want to be as individuals. It is a fluid device to deliver society's arrival to its own most pressing, favoured, altering and transformative destinations.

References

Abdy, M. (ed.) (1999) *Janet and John go Shopping*. London: Section d.

Albery, N. (1994) How to spend millions. *The Guardian*, 21 December.

Alexander, J. (1995) A little flower power can help you slim. *Daily Mail*, 11 February.

Altshuler, B. (1994) *Isamu Noguchi*. London and New York: Abbeville Press.

Argyle, M. (1996) *The Social Psychology of Leisure*. Harmondsworth: Penguin Books.

Balmer, D. (1999a) Just the ticket for no-frills travellers. *The Observer*, Travel News, 26 September.

Balmer, D. (1999b) Tour operators cash in on rush for next eclipse. *The Observer*, 15 August.

Barrett, F. (1995) *The Mail on Sunday*, 22 October.

Barrett, F. (1998) And the Winner is ... Chicken, Prune and Chick Pea Stew. *The Mail on Sunday*, 30 August.

Bauman, Z. (1994) Desert spectacular. In K. Tester (ed.) *The Flâneur*. London and New York: Routledge.

Bennett, C. (1999) Restoration comedy. *The Guardian*, 23 September.

Bevan, S. (1999) National parks plan road tolls. *The Sunday Times*, 21 November.

Boniface, P. (1995) *Managing Quality Cultural Tourism*. London and New York: Routledge.

Boniface, P. (1996) The tourist as a figure in the National Trust landscape. In D.M. Evans, P. Salway and D. Thackray (eds) *The Remains of Distant Times: Archaeology and the National Trust*. Woodbridge: The Boydell Press for The Society of Antiquaries of London and The National Trust.

Boniface, P. (1998a) Are museums putting heritage under the domination of tourism? *Nordisk Museologi* 1, 25–32.

Boniface, P. (1998b) Tourism culture. Research note. *Annals of Tourism Research* 25 (3), 746–9.

Boniface, P. (1999) Meeting needs in heritage cities. In A.P. Russo, A.P. and J. van der Borg (eds) *Proceedings of the International Seminar 'Tourism Management in Heritage Cities'*. Technical Report No 28. UNESCO Venice Office and Technology for Europe (ROSTE).

Boniface, P., Fowler, P. and Stabler, M.J. (2000) *The Centrality of Culture in Tourism: The Causses of France*. CLESC Working paper II: Geographical Paper No. 142. Reading: University of Reading.

Brandt, A. (1997) Denmark's metaphysics of sandcapes. In J. de Graaf with D'Laine Camp (eds) *Europe: Coast Wise: An Anthology of Reflections on Architecture and Tourism*. Rotterdam: 010.

Bray, R. (1999) Takeover fever. *The Guardian*, 7 August.

Burke, J. (1999) Buddha loses his serenity to a theme park. *The Observer*, 5 September.

Carey, K. (1995) *The Third Millennium: Living in the Posthistoric World* (1991, Starseed, The Third Millennium). San Francisco: Harper.

Chatwin, B. (1988) *The Songlines* (1987 Jonathan Cape Ltd). London: Pan Books in association with Jonathan Cape Ltd.

Cochrane, P. (1995) In M. Harrison, *Visions of Heaven and Hell*. London: Channel 4 Television.

Cohen, E. (1995) Contemporary tourism – trends and challenges: Sustainable authenticity or continued post-modernity? In R. Butler and D. Pearce (eds) *Change in Tourism: People, Places, Processes*. London and New York: Routledge.

Commission of the European Communities (1995) Green Paper 'The Role of the Union in the Field of Tourism'. Brussels: Commission of the European Communities, 4 April.

Cossons, N. (1994) *Guide to the Science Museum*. London: Science Museum.

Coupland, D. (1995) *Microserfs*. London: Flamingo.

Crace, J. (1999) Land and freedom. *The Guardian*, 4 May.

DCMS, Tourism Division (1999) *Tomorrow's Tourism: A Growth Industry of the Millennium*. London: DCMS.

de Graaf, J. (1997) Europe: Coast Wise. In J. de Graaf, with D'Laine Camp (eds) *Europe: Coast Wise: An Anthology of Reflections on Architecture and Tourism*. Rotterdam: 010.

Downer, L. (1990) *On the Narrow Road to the Deep North: Journey into a Lost Japan*. London: Sceptre.

Drury, M. (1998) The Director-General's review of the year. *1997/98 Annual Report and Accounts*. London: The National Trust.

Egan, T. (1995) Chairman Gates creates his virtual Xanadu. *The Guardian*, 17 January.

ETB (1998) Brief. *Summer 1998 Newsletter*. London: English Tourist Board.

Few, R. (ed.)(1993) *Caring for the Earth: A Strategy for Survival*. London: Mitchell Beazley.

Forbes, R.J. (1987) The US mature travel market. *Travel and Tourism Analyst*, November.

Frayling, C. (1995) New light on a time of darkness. *The Sunday Times* 10, 21 May.

Gillilan, L. (1995) The house on the hill. *The Guardian* Weekend, 11 February.

Green Globe (1996) *Annual Review 1995/96*. London: Green Globe.

Guixé, M. (1999) Designs (with Bey, J., Jongerius, H. and Wanders, M.). *Couleur Locale*. Rotterdam: 010.

Handy, C. (1994) *The Empty Raincoat: Making Sense of the Future*. London: Hutchinson.

Holland, J. (1994) Castilian village finds the sound of silence deafening. *The European élan*, 18–24 March.

Horne, D. (1992) *The Intelligent Tourist*. McMahons Point: Margaret Gee Publishing.

Humphreys, C. (1990) *Buddhism: An Introduction and Guide* (3rd edn). London: Penguin Books.

Johnson, H. (1997) Foreword. In H. Duijker, *Touring in Wine Country: The Loire*. London: Mitchell Beazley.

Jones, R. and Stuart, L. (1998) Where the fun day now means time well spent. *The Guardian*, 22 August.

Jongerius, H. (1999) Designs (with Bey, J., Guixé, M. and Wanders, M.). *Couleur Locale*. Rotterdam: 010.

Keegan, V. (1999) Sim city for Sibelius. *The Guardian*, 16 September.

Koolhaas, R. (1999) ICA Lecture at the Royal Geographical Society, 28 July.

Lamb, C. (1995) Pop-chart monks snap under the pressure. *The Sunday Times* 1, 15 October.

Lawrence, F. (1999) Are you experienced? *The Guardian*, 12 August.

Laws, E. (1995) *Tourist Destination Management: Issues, Analysis and Policies*. London and New York: Routledge.

Lencek, L. and Bosker, G. (1999). *The Beach: The History of Paradise on Earth*. (1998, New York: Viking Penguin). London: Pimlico.

Lichtenstein, R. and Sinclair, I. (1999) *Rodinsky's Room*. London: Granta Publications.

MacCannell, D. (1999) *The Tourist: A New Theory of the Leisure Class*. Berkeley and Los Angeles: University of California Press.

MacDonald, G.F. and Alsford, S. (1989) *Museum for the Global Village*. Hull (Canada): Canadian Museum of Civilization.

Mazlish, B. (1994) The *flâneur*: From spectator to representation. In K. Tester (ed.) *The Flâneur*. London and New York: Routledge.

McKie, R. (1996) Bike ban shakes New Forest to its roots. *The Observer*, 18 February.

McLuhan, M. (1962) *The Gutenberg Galaxy*. London: Routledge & Kegan Paul.

McLuhan, M. (1967) with V.J. Papanek, J.B. Bessinger, K. Polyani, C.C. Hollis, D. Hogg and J. Jones. *Verbi-Voco-Visual Explorations*. New York: Something Else Press, Inc.

McLuhan, M. (1970) *Counterblast*. London: Rapp & Whiting Ltd.

Midgley, S. (1996) Take a seat and walk the Dales. *The Observer*, 17 March.

Miller, J. and Berry, J. (1999). Goodbye Provence ... *The Sunday Times*, 15 August.

Moulin, C. and Boniface, P. (1999) Routing heritage for tourism: Making heritage and cultural tourism networks for socio-economic development. Paper for the ICOMOS 12th General Assembly and Scientific Symposium, Mexico.

Mulgan, G. (1997) *Connexity: How to Live in a Connected World*. London: Chatto and Windus.

Mumford, L. (1964) *The Highway and the City* (revised edn). London: Secker and Warburg.

National Trust (1995) *Learning from Country Houses*. London: The National Trust.

National Trust (1998) *Volunteering with the National Trust* leaflet. Cirencester: The National Trust.

National Trust (1999a) Caring tourism. *The National Trust Magazine* 88, Autumn. London: The National Trust.

National Trust (1999b) *1998/99 Annual Report to Members*. London: The National Trust.

Nicholson-Lord, D. (1995) 'Greys' set to dominate the leisure market. *The Independent*, 27 January.

Page, S. (1995) *Urban Tourism*. London and New York: Routledge.

Parks Canada (1994) *Guiding Principles and Operational Policies*. Ottawa: Minister of Supply and Services Canada.

Paton Walsh, N. (1999) Escape guide: Travel on the internet. *The Observer*, 18 April.

Pattullo, P. (1996) *Last Resorts: The Cost of Tourism in the Caribbean*. London: Cassell and Latin America Bureau (Research and Action) Ltd.

Pearman, H. (1997) Rebuilt just as they like it. *The Sunday Times*, 23 April.

Penman, A. (1994) How a million selling hit record shattered the lives of the monks of Santo Domingo. *The Mail on Sunday*, 5 June.

Peters, T. (1994) *The Tom Peters Seminar: Crazy Times Call for Crazy Organizations*. London: Macmillan.

Phillips, I. (1999) The future's Orange. *The Observer* Magazine, 11 July.

Pierce, J. (1995) Going green. *American Vogue*, January.

Pietrasik, A. (1999) Seaside restoration. *The Guardian*, Travelogue, 7 August.

Pine, B.J. and Gilmore, J. (1999) *The Experience Economy: Work is Theatre and Every Business a Stage*. Cambridge, MA; Harvard Business School Press.

Plog, S.C. (1994) An extraordinary industry faces superordinary problems. In W. Theobold (ed.) *Global Tourism: The Next Decade*. Oxford: Butterworth-Heinemann.

Powerhouse Museum (1995) *Ken Done: The Art of Design*. Haymarket: Powerhouse Press.

Raban J. (1974) *Soft City*. London: Fontana/Collins.

Robinson, M. and Boniface, P. (eds) (1999) *Tourism and Cultural Conflicts*. Wallingford: CABI Publishing.

Rogers, R. (1995) Looking forward to Compact City. *The Independent*, 20 February.

Rogers, R. (1997) *Cities for a Small Planet*. P. Gumuchdjian (ed.). London: Faber and Faber.

Ross, V. (1995) Canadian's work at modern art museum. *The Globe and Mail*, 29 September.

Rowinski, P. (1995) The wall moves over the ocean. *The European* Magazine, 23–29 June.

Siebert, C. (1995) Composting in paradise. *American Vogue*, January.

Shields, R. (1994) Fancy footwork: Walter Benjamin's notes on *flânerie*. In K. Tester (ed.) *The Flâneur*. London and New York: Routledge.

Sage, A. (1995) Green revolution brings new dawn to doomed village. *The Observer*, 19 February.

Spencer. M. (1995) Beyond the fringe. *Vogue*, January.

Street, J. (1995) On foot into the green woods. *The European* Magazine, 23–29 June.

Sudjic, D. (1995) They came, they saw, they ate pizza. *The Guardian* Arts, 4 August.

Suzuki, D.T. (1993) *An Introduction to Zen Buddhism*. C. Humphreys (ed.). London: Rider Books.

Taylor, B. (1999) Monk turns a pig farm into a garden outpost of Japan. *Daily Mail*, 23 April.

Tolhurst, C. (1994) Crystals, cults & communes: Australia comes of New Age. *Australia Unplugged: Escape & Discover Down Under*. Brochure. Sydney: Australia Tourist Commission.

Turner, L. and Ash, J. (1975) *The Golden Hordes: International Tourism and the Pleasure Periphery*. London: Constable.

Urry, J. (1995) *Consuming Places*. London and New York : Routledge.

van Harssel (1994) The senior travel market: Distinct, diverse, demanding. In W. Theobold (ed.) *Global Tourism: The Next Decade*. Oxford: Butterworth-Heinemann.

van Meggelen, B. (1999) *Rotterdam is Many Cities*. NR.1 February. Rotterdam Cultural Capital 2001.

Victoria & Albert Museum (1991) Visions of Japan exhibition leaflet.

Wanders, M. (1999) Designs (with Bey, J., Guixé, M. and Jongerius, H.). *Couleur Locale*. Rotterdam: 010.

Ward, D. (1995) Peak rate scheme for park motorists. *The Guardian*, 10 March.

Weiss, T. (1999) Restoration, revival and innovation. *Couleur Locale*. Rotterdam: 010.

WTO (1991) *Tourism to the Year 2000: Qualitative Aspects Affecting Global Growth*. Madrid: WTO.

WTO (1994) *National and Regional Tourism Planning*. London and New York: Routledge.

WTO (1995) *WTO News* No 8, December.

WTO (1998). Baltic Sea destinations poised for tourism growth. Press release from World Tourism Organization.

Zukin, S. (1996) Space and symbols in an age of decline. In A.D. King (ed.) *Re-Presenting the City: Ethnicity, Capital and Culture in the Twenty-First Century Metropolis*. London: Macmillan Press Ltd.

Index

Individual place or item names are often used, but instead the features may appear under their country or category; references to countries alone relate to general aspects

185